Smart Bioenergy

Daniela Thrän
Editor

Smart Bioenergy

Technologies and concepts for a more flexible bioenergy provision in future energy systems

Springer

Editor
Daniela Thrän
Department of Bioenergy
Helmholtz Centre for Environmental
Research – UFZ
Leipzig, Germany

Deutsches
Biomasseforschungszentrum – DBFZ
Leipzig, Germany

Bioenergy Systems
University of Leipzig
Leipzig, Germany

ISBN 978-3-319-16192-1 ISBN 978-3-319-16193-8 (eBook)
DOI 10.1007/978-3-319-16193-8

Library of Congress Control Number: 2015938445

Springer Cham Heidelberg New York Dordrecht London

Printed on acid-free paper

Springer International Publishing AG Switzerland is part of Springer Science+Business Media (www.springer.com)

Contents

Contributors

Eric Billig Deutsches Biomasseforschungszentrum GmbH – DBFZ, Leipzig, Germany

Jaqueline Daniel-Gromke Deutsches Biomasseforschungszentrum GmbH – DBFZ, Leipzig, Germany

Subhashree Das Department of Bioenergy, Helmholtz Centre for Environmental Research – UFZ, Leipzig, Germany

Martin Dotzauer Deutsches Biomasseforschungszentrum GmbH – DBFZ, Leipzig, Germany

Marcus Eichhorn Department of Bioenergy, Helmholtz Centre for Environmental Research – UFZ, Leipzig, Germany

Fabian Jacobi Deutsches Biomasseforschungszentrum GmbH – DBFZ, Leipzig, Germany

Jan Khalsa Deutsches Biomasseforschungszentrum GmbH – DBFZ, Leipzig, Germany

Claudia Kirsten Deutsches Biomasseforschungszentrum GmbH – DBFZ, Leipzig, Germany

Marco Klemm Deutsches Biomasseforschungszentrum GmbH – DBFZ, Leipzig, Germany

Volker Lenz Deutsches Biomasseforschungszentrum gGmbH – DBFZ, Leipzig, Germany

Jan Liebetrau Deutsches Biomasseforschungszentrum GmbH – DBFZ, Leipzig, Germany

Stefan Majer Deutsches Biomasseforschungszentrum GmbH – DBFZ, Leipzig, Germany

Franziska Müller-Langer Deutsches Biomasseforschungszentrum GmbH – DBFZ, Leipzig, Germany

Karin Naumann Deutsches Biomasseforschungszentrum GmbH – DBFZ, Leipzig, Germany

Andreas Ortwein Deutsches Biomasseforschungszentrum gGmbH – DBFZ, Leipzig, Germany

Philip Tafarte Department of Bioenergy, Helmholtz Centre for Environmental Research – UFZ, Leipzig, Germany

Daniela Thrän Department of Bioenergy, Helmholtz Centre for Environmental Research – UFZ, Leipzig, Germany

Deutsches Biomasseforschungszentrum – DBFZ, Leipzig, Germany

Bioenergy Systems, University of Leipzig, Leipzig, Germany

Janet Witt Deutsches Biomasseforschungszentrum GmbH – DBFZ, Leipzig, Germany

Chapter 1
Introduction

Daniela Thrän

Abstract Biomass is the most relevant renewable energy source, with a wide range of possible and established methods to apply biomass for energy generation. In regards to the supply of sustainable energy, not only the provision of technology but also the integration of this technology into the system will be considerably important. This demands a change in bioenergy provision which is comparable to the transition from traditional to modern biomass use. The need for further development in the provision of bioenergy is also underlined by the challenges affecting the biomass resource base, including increasing demands for biomass for food, feed, materials and fuel. Furthermore, this is underlined by the major concerns surrounding factors relating to the land such as soil, nutrients and biodiversity.

Germany has implemented an active policy for the transition of the energy system towards greater use of renewable energy sources more than a decade ago, which has led to a strong increase in the amount of biomass used for electricity, heat and the provision of transport fuel. With relevant shares of electricity from wind and solar the need for better system integration is on the agenda. The situation in Germany can therefore provide interesting insights into the challenges and opportunities of using bioenergy in its new role. This will be elaborated on step by step in this book, starting with issues relating to the market and resource base, then moving on to analysis of the technical options, followed by the modeling of the effects on the German energy system in a case study and in conclusion focusing on the most promising fields as well as the missing elements for a successful transition.

D. Thrän (✉)
Department of Bioenergy, Helmholtz Centre for Environmental Research – UFZ, Permoset Straße 15, 04318 Leipzig, Germany

Deutsches Biomasseforschungszentrum – DBFZ, Torgauer Straße 116, 04347 Leipzig, Germany

Bioenergy Systems, University of Leipzig, Grimmaische Straße 12, 04109 Leipzig, Germany
e-mail: daniela.thraen@ufz.de

D. Thrän (ed.), *Smart Bioenergy*, DOI 10.1007/978-3-319-16193-8_1

1.1 Bioenergy Today

The transition of the energy system towards greater use of renewable energy is a precondition for the envisaged reduction of greenhouse gas emission [12], for a sustainable use of the finite resources [4] and for a payable and fair access to energy [1]. The use of renewable energy sources has been increased worldwide over the last decade with biomass as the most important source covering 10 % of the total global primary energy demand [17].

Biomass is a renewable carbon source which uses the process of photosynthesis to generate hydrocarbons. The most relevant biomass resources for bioenergy are energy crops, residues and by-products from forestry, agriculture, the wood and food processing industry, residues from gardening and landscape management as well as organic waste from industry and final consumers. In general, all these renewable hydrocarbon sources can be used to substitute all fossil fuels, including natural gas, liquid fuels and coal. In practice, the effort expenditure needed for this substitution differs greatly due to different chemical structures, inorganic compounds and trace elements in the biobased materials, which affect the conversion unit and can cause corrosion, slagging, lower efficiency and higher emissions [14].

The use of biomass is realized through a number of multi-step pathways, including resource provision, transportation, storage and condition, conversion of the biomass into biofuel and finally conversion into useable energy (Fig. 1.1).

There are thermo-chemical, bio-chemical and physico-chemical conversion systems available to produce solid, liquid and gaseous fuels from biomass. These fuel sources are then combusted to generate heat and power in stationary and mobile applications:

- Thermo-chemical conversion involves the use of a system to transform solid biomass into charcoal, pyrolysis oil, product gas and other intermediates (thermochemically treated solid biomass). The next step sees these intermediates be transformed into dedicated bioenergy carriers (synthetic biofuels, synthetic natural gas (SNG)).
- Physico-chemical conversion is used for oil resources to provide vegetable oil or biodiesel
- Biochemical conversion includes the anaerobic digestion of sugar and starch to produce biogas and the fermentation of sources containing specific sugars to bioethanol.
- All the converted biofuels can be processed further, i.e. product gas to liquid biofuels ("bio-to-liquid"), vegetable oil to hydrogenated biofuels ('HVO'), so in theory pathways are possible from almost every resource to every energy carrier. In practice, these concepts are still however at the research and demonstration stage.

Today, biomass contributes to the heat and power market as well as for the provision of renewable transportation fuels [17]. Figure 1.2 gives an overview of the relevance of the different fields of application worldwide. The provision of heat through biomass is the most important field. Traditional use of biomass to produce heat requires the highest share of resources. Even if realistic figures of the total use

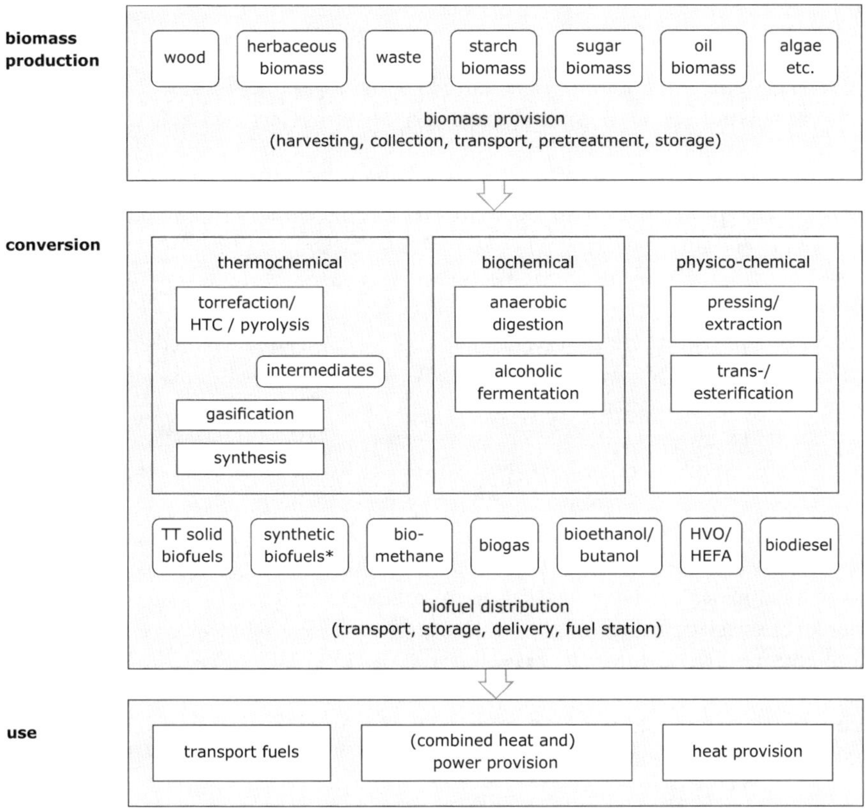

Fig. 1.1 Pathways of technologies for biomass feedstock conversion (*TT* thermochemically treated; * Fischer-Tropsch Diesel, DME, Biohydrogen etc.)

Fig. 1.2 Global biomass flows in 2012 (Primary energy from [17], secondary energy from [3, 17] and own calculation)

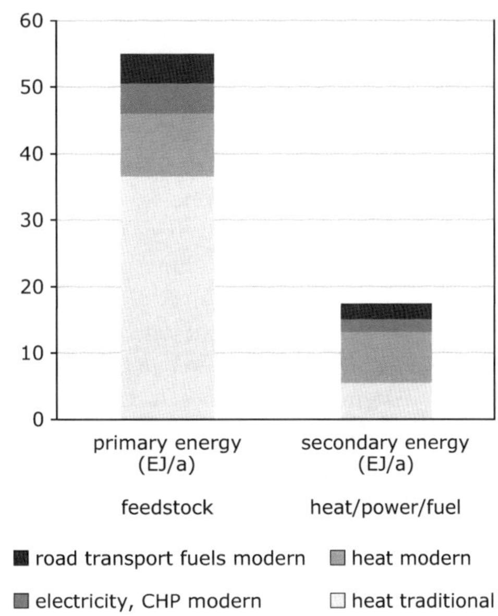

of biomass are uncertain due to additional activities in the informal sector, which are normally not reported in the statistics [9], resources are used rather inefficiently in this sector, in comparison: open fire places and stoves for heat provision in the "traditional bioenergy sector" are characterized by conversion rates of 10–20 % [6]. This kind of application is still the major conversion pathway for biomass feedstock to provide energy in the global context (about 35 EJ/a; see Fig. 1.2). Modern biomass uses like, for example, biofuel provisions for transport, combined heat and power provision and modern heat provision in boilers and stoves with a higher energy efficiency rate are applied to about 20 EJ/a of the biomass feedstock. Because of the much higher conversion efficiency of modern biomass, it is able to deliver almost 70 % of the secondary energy.

Electricity, modern combined heat and power (CHP) and biofuels for the transport sector provide a minor contribution, and are used especially in countries committed to the use of renewable energy. Today, these facilities often run for the whole year on full load, producing biofuels or bioenergy regardless of the actual energy demand and supply situation. The transition of the energy system will change this picture dramatically: the increasing use of solar and wind energy requires the provision of electricity from biomass in periods of insufficient wind speed or insufficient sunlight; the use of different renewable sources in combination with one another can also occur in the heating sector, for example the combined use of biomass systems and solar collectors, while in the field of biofuels the combined provision of biofuels and other biomass products will be of higher interest.

1.2 The Way Forward: Traditional, Modern and Integrated Bioenergy Provision

The global figures relating to current bioenergy provision clearly indicate that the role of bioenergy and the transition processes towards renewable energy use in different regions of the world vary. The transition from traditional biomass use to modern biomass use guarantees more energy efficiency and can in most cases reduce local emissions [19]. Another transition taking place is that from modern biomass use as a stand-alone concept to biomass having an integrated role in an energy system mainly based on renewable energy sources. The expected new way for bioenergy use is shown in Fig. 1.3. For biomass to have an integrated role, two questions must be addressed: The first question is how to provide bioenergy with efficient technologies. The second question is, which kind of bioenergy is able to best support demands for a secure, cost efficient and climate protective energy system. The provision of flexible bioenergy requires the conversion system to have a greater capacity and a greater number of control units. In many cases, this would mean reducing the full load production of a conversion unit from basic load (8,000 h per year full load) to part load or seasonal operation, or to change the technical concepts to smaller conversion units to be operated in modules. These additional measures need to be beneficial to the overall energy supply, for this reason

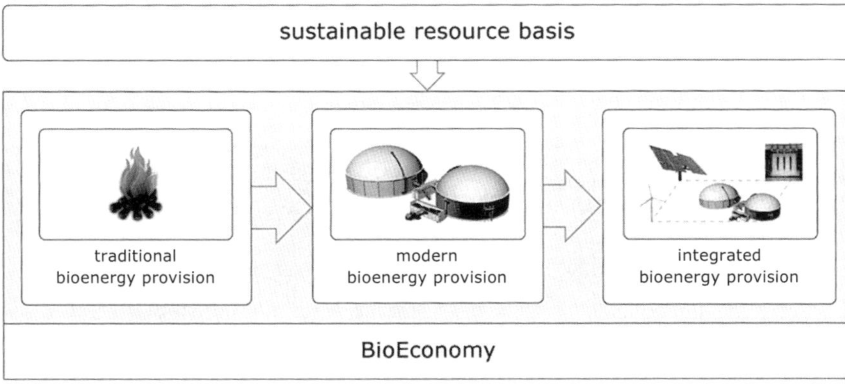

Fig. 1.3 The development of bioenergy as one column of the transition of the energy system

promising solutions can be expected to be different in different countries, i.e. Sweden as a country with high hydropower potential focuses much more on the use of biomass in the transport sector.

Not only demand but also supply is characterised by major changes. In regards to future systems, it is expected that the demand for biomass will increase – for food and feed but also for materials – while different factors will reduce the resource base. Additionally, there is an increasing concern that the direct and indirect effects on bioenergy provision could contradict the intended effect, for example change could occur as a result of energy crop production through direct and indirect land use, through airborne emissions of greenhouse gas relevant substances (methane, nitrogen dioxide) along the process chain, through environmental effects due to the increasing use of by-products from agriculture and forestry [15, 18]. On the other hand, additional potentials on land and biomass are discussed, and there is the chance of increasing the use of marginal land, if the use of biomass is embedded in an integrated approach for sustainable land use [5] (Fig. 1.3). Finally, the major part of the resource basis is provided and used on a local and regional scale; trans-regional transport and trade is only suitable for specific biofuels, such as wood pellets, biodiesel and bioethanol [13].

1.3 The German Transformation: Demanding a New Role for Bioenergy?

Germany has implemented an active policy for the transition of the energy system towards the use of renewable energy more than a decade ago. The development of the renewable energy sector in Germany actively started to gain momentum in 1991 when the electricity feed-in law came into force, which was renamed as the "EEG"– Renewable Energy Sources Act in 2000 and has since undergone several amendments [7, 20]. The EEG aims to enforce technological development in order to introduce renewable energy into the electricity market and integrate it into the energy system. Since 2001, the European energy policy has promoted electricity production from renewable energy sources on the domestic electricity market [8]. Currently 20.3 % of the electricity consumed in Germany is produced from renewable energy that is mainly supported by the EEG, with the fluctuating use of wind and photovoltaic power accounting for one half, and biomass and hydropower accounting for the other [2].

A biofuel policy was implemented in 2003 in the transport sector, introducing a tax exemption for biofuels and transforming the system into a quota system from 2007 onwards [11]. This led to a share of biofuels in the road transport sector of 6.25 % since 2010. Up to now, all renewable energy sources for the provision of the transport sector are derived from biomass [16].

Furthermore, biomass is dominant in the heat sector, accounting for 90 % of renewable heat produced. The heat sector is much more market driven. An increase of biomass use occurred from 2003 due to the consumer reaction to increased prices of fossil fuels. In conclusion, 65 % of the renewable energy provided in Germany is a product of biomass, with the potential to reduce 65 million tons of CO_2-emissions, which is equivalent to the total amount of greenhouse gases emitted in 2012 [2].

In many respects, Germany is extreme in exploring the new role of bioenergy. It is a densely populated nation with a high energy demand, has a strong agricultural sector and limited access to other continuous renewable resources such as hydro or geothermal energy. The International Renewable Energy Agency (IRENA) has ranked Germany in first position in the categories "*Renewable power per capita (not incl. hydro)*", "*total installation of PV*" and "*total installation of wind power*" and with place 3 in the categories "*total installations of bio-power*" and "*total installations of biodiesel production*" in the REN 21 report 2013 [17].

The development of the biomass input of the different energy carriers in Germany is shown in Fig. 1.4. Germany's renewable energy policy led to a strong increase of biomass use for energy provision in all three sectors over the last decade. Today more than 50 % of residues and the waste stream is used, and energy crops are cultivated on around 2 million ha of arable land [10]. The further additional biomass potential and expansion of the resource base is limited.

For the analysis of the possible new role of bioenergy in future energy systems these frame conditions offer interesting insights. This is the reason Germany has been selected for case study to demonstrate the possible options for a more system oriented provision of bioenergy.

1.4 Set-Up of the Book

In this book the possible new roles of bioenergy are described from a technical perspective while also taking into consideration market expectations. The aim is to find out what are the most promising solutions in the short- and midterm because

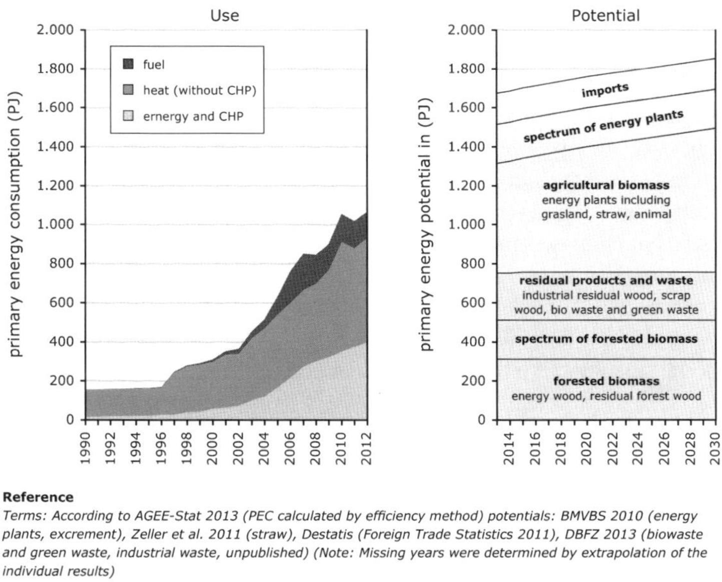

Fig. 1.4 Use of biomass for energy provision in Germany – historical development and expected availability of biomass

increasing flexibility always requires higher effort expenditure to produce certain amounts of energy.

The expected demand from future energy systems is the starting point for this investigation. This includes the question what kind of energy is necessary (power, heat, transport fuel etc.) and under which frame conditions the demand is expected (continuous versus discontinuous demand). In Chap. 2 the different energy markets are therefore analysed, including the status quo of biomass use and expected changes as well as market demands which are resulting from the process of transition of Germany's energy system. Germany is taken as a case study because of the extreme frame conditions for demand orientated provision. Furthermore, there is good data available for comparison as well as some practical experience in this field.

The second big driving force for a change in the role of bioenergy in future systems is the expected biomass potentials under the currently discussed sustainability criteria. This is analyzed in Chap. 3. Due to the complex international interdependencies of biomass potentials, the analysis of the resource base is provided on a global basis, analyzing the relevant driving forces and frame condition and to discuss with them the future availability of biomass.

The options for new concepts and technologies are investigated in the following chapters, referring to power provision from solid biofuels (Chap. 4), power provision from biogas (Chap. 5), heat provision from solid fuels (Chap. 6) and biofuels for the transport sector (Chap. 7). Additionally, the concepts and technologies to upgrade biomass to a better defined intermediate fuel which can be used in more complex conversion technologies are provided in Chap. 8.

Finally, a case study of a German region examines the effects of system optimized power provision from different renewable energy sources with bioenergy as a flexible option to balance the regional power supply system (Chap. 9).

From today's perspective it is very difficult to draw a full picture of the future application of bioenergy. However, the comprehensive analysis of the new challenges of the different markets and the potential technical answers could reveal the options for the transition of bioenergy use in future energy systems with higher share of renewables. This will be summarized in Chap. 10.

A final remark: biomass is a limited resource and at the moment how to satisfy the future demands for energy and materials is questionable; the reduction of the overall demand – by efficiency and sufficiency approaches – seems to be a precondition for the sustainable integration of bioenergy in future systems.

References

1. AGECC, *Summary Report and Recommendations* (AGECC, New York, 2010)
2. BMU, *Zeitreihen zur Entwicklung der erneuerbaren Energien in Deutschland, unter Verwendung von Daten der Arbeitsgemeinschaft Erneuerbare-Energien-Statistik (AGEE-Stat)* (BMU, Berlin, 2012)
3. BP Energy Outlook 2035. http://www.bp.com/en/global/corporate/about-bp/energy-economics/energy-outlook/outlook-to-2035.html. Retrieved 7 Apr 2014
4. S. Bringezu, H. Schütz, W. Pengue, M. O'Brien, F. Garcia, R. Sims, R. Howarth, L. Kauppi, M. Swilling; U. A.: UNEP, *Assessing Global Land Use: Balancing Consumption with Sustainable Supply. A Report of the Working Group on Land and Soils of the International Resource Panel* (2014). ISBN 978-92-807-3330-3
5. J. Dauber, C. Brown, A.L. Fernando, J. Finnan, E. Krasuska, J. Ponitka, D. Styles, D. Thrän, K.J. Van Groenigen et al., Bioenergy from "surplus" land: environmental and socio-economic implications. BioRisk **7**, 5–50 (2012)
6. O. Edenhofer, R. Pichs-Madruga, Y. Sokona, K. Seyboth, P. Matschoss, S. Kadner, T. Zwickel, P. Eickemeier, G. Hansen; U. A.: IPCC, *IPCC Special Report on Renewable Energy Sources and Climate Change Mitigation. Prepared by Working Group III of the Intergovernmental Panel on Climate Change* (Cambridge University Press, Cambridge, UK/New York, 2011). ISBN 978-1-107-02340-6
7. EEG, Gesetz für den Vorrang Erneuerbarer Energien (Erneuerbare-Energien-Gesetz – EEG) Konsolidierte (unverbindliche) Fassung des Gesetzestextes in der ab 1. Januar 2012 geltenden Fassung. Renewable Energy Resource Act of Germany, 2012
8. Eurostat (2010) European Commission, Renewable Energy Statistics
9. A. Faaij, M. Junginger, C.S. Goh, A general introduction to international bioenergy trade, in *International Bioenergy Trade – History, Status & Outlook on Securing Sustainable Bioenergy Supply, Demand and Markets* (Springer, Dordrecht/Heidelberg/New York/London, 2014)
10. Fachagentur Nachwachsende Rohstoffe (FNR), *Anbau nachwachsender Rohstoffe 2012 auf 2,5 Millionen Hektar* (FNR, Gülzow, 2012)
11. Gesetz zur Einführung einer Biokraftstoffquote durch Änderung des Bundes-Immissionsschutzgesetzes und zur Änderung energie- und stromsteuerrechtlicher Vorschriften vom 18 Dec 2006
12. N. Nakicenovic, A. Grübler, H. Ishitani, T. Johansson, G. Marland, J.R. Moreira, H.-H. Rogner, *Climate Change 1995. Second Assessment Report. Chapter B Energy Primer* (Intergovernmental Panel on Climate Change, UNDP, 1995)

13. M. Junginger, C.S. Goh, A. Faaij, *International Bioenergy Trade – History, Status & Outlook on Securing Sustainable Bioenergy Supply, Demand and Markets* (Springer, Dordrecht/Heidelberg/New York/London, 2014). ISBN 978-94-007-6981-6
14. M. Kaltschmitt, H. Hartmann, H. Hofbauer, *Energie aus Biomasse: Grundlagen, Techniken und Verfahren* (Springer, Berlin, 2009). ISBN 9783540850946
15. J. Liebetrau, J. Clemens, C. Cuhls, C. Hafermann, C. Friehe, P. Weiland, J. Daniel-Gromke, Methane emissions from biogas producing facilities within the agricultural sector. Eng. Life Sci. **10**, 595–599 (2010)
16. K. Naumann, K. Oehmichen, M. Zeymer, K. Meisel, *Monitoring Biokraftstoffsektor* (Nr. DBFZ Report Nr. 11 (2. Auflage)). (DBFZ Deutsches Biomasseforschungszentrum gGmbH, 2014)
17. REN21, *Renewables 2013 Global Status Report* (REN21 Secretariat, Paris, 2013). ISBN 978-3-9815934-0-2
18. T. Searchinger, R. Heimlich, R.A. Houghton, F. Dong, A. Elobeid, J. Fabiosa, S. Tokgoz, D. Hayes, T. Yu, Use of U.S. croplands for biofuels increases greenhouse gases through emissions from land-use change. Science **319**, 1238–1240 (2008)
19. K.R. Smith, Health impacts of household fuelwood use in developing countries. Unasylva **57**, 41–44 (FAO, Rome, 2006)
20. Stromeinspeisegesetz (1991) Electricity Feed-in Law of Germany

Chapter 2
Demand for the Flexible Provision of Bioenergy Carriers: An Overview of the Different Energy Sectors in Germany

Martin Dotzauer, Karin Naumann, Eric Billig, and Daniela Thrän

Abstract Today bioenergy is the most important renewable energy carrier in Germany and yet it is mostly provided constantly as a base load, for example most biogas plants are in constant production at full load. In numerous ways, bioenergy could be a flexible option to satisfy the fluctuating demand for electricity, heat and transport fuels. In the power sector, biomass is a short-term option to meet the increasing need for flexible power generation, while wind or solar power are characterized by an alternating feed-in. Biogas plants in particular are ideal for providing power on demand for a stable electricity provision with a high percentage from renewables. The heat sector is well established for heat only provision, but has to integrate future combined heat and power concepts. Therefore, an optimal alignment between heat and power generation is required, if a high overall performance is to be achieved. Furthermore, the general decrease in heat through efficiency measures will change the way in which heat will be generated with more versatile load curves where flexible energy provision is favoured. In the transport sector flexibility is necessary in the form of a varying feedstock basis for consistent liquid biofuel products. For example, bioethanol could be made from sugar beets or cereals, where sugar, or starch converted into sugar, is processed by fermentation into alcohol. Second generation bioethanol is based on cellulose enzymatic split into single sugar molecules. Additionally, biomethane as a potential substitute for natural gas can be applied in different sectors and is predestined for flexible energy provision. A local and temporal decoupling of energy source generation, the well-established gas grid

M. Dotzauer (✉) • K. Naumann • E. Billig
Deutsches Biomasseforschungszentrum GmbH – DBFZ,
Torgauer Str. 116, 04347 Leipzig, Germany
e-mail: martin.dotzauer@dbfz.de; karin.naumann@dbfz.de; eric.billig@dbfz.de

D. Thrän
Department of Bioenergy, Helmholtz Centre for Environmental Research – UFZ,
Permoset Straße 15, 04318 Leipzig, Germany

Deutsches Biomasseforschungszentrum – DBFZ, Torgauer Straße 116,
04347 Leipzig, Germany

Bioenergy Systems, University of Leipzig, Grimmaische Straße 12,
04109 Leipzig, Germany
e-mail: daniela.thraen@ufz.de

D. Thrän (ed.), *Smart Bioenergy*, DOI 10.1007/978-3-319-16193-8_2

and the interchangeability with natural gas are all aspects that support this. It is expected that the different markets for power, heat and fuels will be more closely linked by the mid term. Here, some additional combinations of bioenergy with other renewables (i.e. power-to-gas) can provide flexible energy in different sectors additionally.

2.1 Introduction

Since the use of renewable energy sources has been recognized as playing a major role in counteracting climate change, in many countries the direction and design of the energy policy framework has been adapted in favor of promoting renewable energies, leading to ambitious targets for renewables to account for a larger share in the overall energy mix. Furthermore, in terms of the provision, use and integration of bioenergy, the specific political framework conditions are very relevant. Today, biomass is applied for the provision of electricity, heat and transport (see Chap. 1). Bioenergy is already highly relevant in all three final energy sectors and in terms of the transition targets for energy systems, the different markets are affected in different ways. Additionally, the substitution of natural gas with biomethane is of increasing interest, because it enables the renewable carbon carrier to be applied in many different fields. In this chapter, the four energy markets (electricity, heat, transport fuels and biomethane) are described, including the regulatory frameworks, their significance for renewable energy provision, the spatial impacts on particular markets as well as the necessary technical requirements.

The analysis focuses on the market situation in Germany, also including the European conditions. Germany is a country with a high population density and a high energy demand from different industries. In 2012 the total primary energy consumption in Germany amounted to 13,757 PJ. Germany has very little fossil energy resources of its own. The most important sources for energy provision are mineral oil (33.0 %) and natural gas (21.5 %), which are mainly imported. Between 1990 and 2012, the share of renewable energy within the German energy system has increased more than tenfold – from 1.3 % to 11.6 % of primary energy consumption

Table 2.1 Status quo and targets for the transition towards renewable energies and bioenergy in Germany (Based on [1] and [2])

	Status of renewables in 2010/2012	Renewable energy target 2020	Bioenergy target 2020
	Share of renewables in the final energy demand (%)	Share of renewables in the final energy demand (%)	Share of bioenergy in the renewable energy section (%)
Electricity generation	16.8/22.6	38.6	22.8
Heating and cooling	9.8	15.5	78.7
Transport	5.8	13.2	Almost 100

[3]. The German national renewable energy action plan (nREAP) of 2010 includes commitments to targets for renewable energies and for bioenergy in various energy sectors by 2020 (Table 2.1). By 2020 the share of biomass is expected to account for almost 10 % of the total final energy consumption (8,859 PJ) in Germany [4]. Thereby, biomass shall contribute with 22.8 % to the electricity sector, 78.7 % to the heating and cooling sector and almost exclusively provide for the transportation sector with almost 100 % [2].

2.2 Electricity Market

2.2.1 The Political Framework

The provision of electricity from renewables is backed up by Europe's Renewable Energy Directive [5], which sets targets for renewable energy in gross final electricity consumption of at least 20 % by 2020. National targets are not specified there, but the methodology to create national energy action plans describes how to determine overall targets in different sectors.

The national regulatory framework for the provision of renewable electrical energy in Germany is first of all determined by the Renewable Energy Sources Act ("Erneuerbare Energien Gesetz" – EEG) [6]. The EEG Act first became effective in 2000. The EEG Act supports the provision of electricity from biomass in "bioenergy-only"-plants with a power capacity of up to 20 MW. The two main features are the feed-in priority for electricity from renewable power plants and a system of guaranteed feed-in tariffs for different technologies as well several plant sizes. Both aspects ensure that renewable electricity has an advantage over power from nuclear and fossil fuels and thus strongly support market access. The reward system for biomass plants considers the year that operations started, the scale of the conversion plant and the type of biomass used for production. Furthermore, there are several kinds of additional rewards, for example for upgrading from biogas to biomethane, for coupling out of heat for external use and for innovative technologies, which have been integrated by amendments in 2004, 2009 and 2012.

With the amendment of the EEG in 2012 a more market-oriented operation of power provision from biomass was targeted. Therefore, a new reward concept was implemented to support the direct marketing of electricity through the market bonus. The market bonus counterbalances the distribution between spot market prices and the general fixed reward. It is set to the difference between the average monthly price and the fixed reward. Under direct marketing a management bonus is provided as an incentive and to compensate the distribution costs. Additionally, biogas plants can receive a flexibility bonus, when they have the ability to shift the power feed in. The flexibility bonus enables plants to operate below full capacity, to regulate their power according to fluctuant prices on spot markets as well as fluctuating demand [7].

2.2.2 The Market Situation

In 2012 the share of renewable energies in the electricity market grew to account for 22.6 % of Germany's energy mix as shown in Fig. 2.1. Although bioenergy has a relatively small amount of installed capacity (10 %) among the renewable energies compared to wind or solar power, due to its high full load hours (about 7,500) it has a relatively high percentage (31 %) of overall renewable power feed-in.

The provision of renewable power generation from biomass includes electricity generation from biogas, solid fuels and liquid fuels and is carried out in combined heat and power installations as well as in electricity-only installations. The co-combustion of biomass in fossil-fuel plants is not supported and therefore only plays a minor role. The generation from biogas accounts for almost two thirds of all installed capacity and even more of the produced energy (Table 2.2). Biogas plants are evenly distributed across Germany, many of which are representative of the average plant size (0.41 MW). Here, decentralised combined heat and power units using biomethane have been included. A greater focus on the biomethane market on the whole will be provided in Sect. 2.5.

Solid fuel plants provide around one third of the installed capacity and a little less of the produced energy. The average plant size of 2.9 MW results from a large variety of installed capacities in the portfolio of plants.

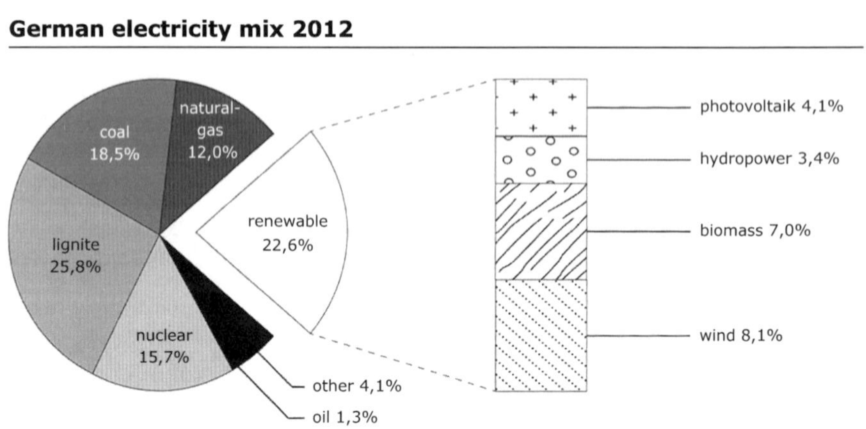

Fig. 2.1 Germany's electricity mix in 2012, share of power produced (Source: Agency for Renewable Energies [8])

Table 2.2 Biomass plants' statistics 2012 [9]

	Count	Capacity	Average size	Feed-in
	[n]	[MW]	[MW]	[TWh]
Biogas	7,400	3.1	0.41	23.1
Solid fuels	540	1.6	2.9	8.4
Liquid fuels	1,050	0.17	0.16	0.25

Even though liquid biomass could be operated as flexible as or even more flexibly than biogas, it has not played such an important role in the recent development of flexible bioenergy provision. A global perspective of the liquid biomass sector is given in Sect. 2.4.

2.2.3 Future Markets for Electricity from Biomass

The transition of power generation from more renewables also leads to different supply patterns: Historically speaking, the electricity supply in Germany could be divided into the base load supply from power plants running 8,000 h per year and the peak load supply that was only added at times of higher demand. The highest demand for power typically occurred during the daytime in winter, peaking in the late afternoon (Fig. 2.2, left). Typically, the base load supply was provided by coal and nuclear power plants, whereas the peak load supply came from natural gas installations. With the increasing supply from wind and solar power, the supply pattern changed to variable proportions of wind and solar power and a remaining demand to be covered by flexible and controllable power plants – the so-called residual load. Due to the sunlight dependency of solar power, the residual load is at a minimum during noon (Fig. 2.2, right), and can also become negative on sunny day, when the demand for power is generally low (i.e. on Sundays or holidays). Future markets for renewable electricity will therefore have to focus on an efficient supply of the residual load from biomass and other flexible provision options.

So far biomass plants have been running 8,000 h per year, but within the announced change in renewable power provision from the feed-in priority with fixed tariffs to a more market-oriented model, the question is really which markets would

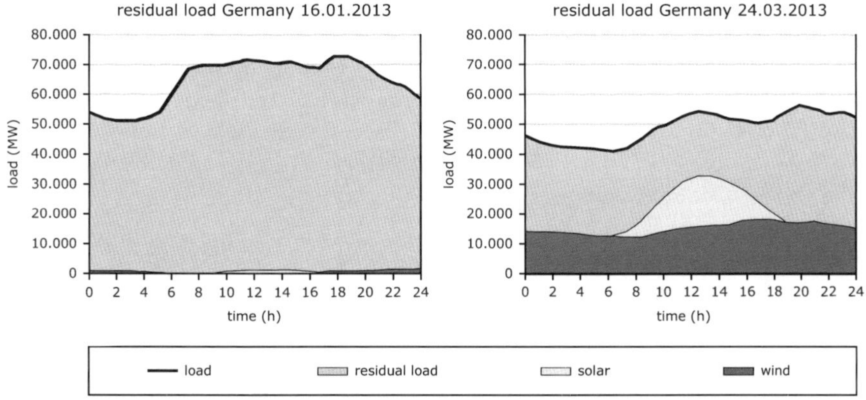

Fig. 2.2 Recent residual loads in Germany under different consumption and renewable feed-in patterns [10]

be most suitable for the different biomass plants. Electricity markets are sub-divided into energy-only markets for the trade in energy and markets for control reserve (see also: https://www.entsoe.eu/fileadmin/user_upload/_library/publications/entsoe/Operation_Handbook/Policy_1_final.pdf and https://www.regelleistung.net/ip/action/static/marketinfo)

Energy-Only Market Electricity markets in Central Europe and Germany can be subdivided into electricity that is traded "over the counter" (OTC) or through power exchanges that are organised markets (PXs). There are two main segments in power exchanges for electricity. The future market, for advance trading for up to 6 years ahead is located at the European Energy Exchange (EEX) in Leipzig [11]. The relevant spot-market for short-term trading is located at the European Power Exchange (EPEX) in Paris. For a flexible provision of electrical energy, the relevant market is European Power Exchange, where electricity is traded within a one-price auction. The order is structured by individual hours for day-ahead trading [12]. There is also an intraday market with quarters of hours, but normally this is too quick for biogas plants to react to price signals and to schedule their production accordingly. To participate in these markets a certain amount of power has to be provided.

Markets for Control Reserve Besides the energy supply, some electricity is also needed to provide a secure power-supply infrastructure. For network operation, control reserve is necessary to balance forecast errors for power generation and consumption, because of the need for an anytime equilibrium of feed-in and the delivery of electrical energy. Control reserve is needed at the top level of network topology in the distribution network. In Germany, there are four distribution network operators that enable the demand for control reserve to accumulate in an announcement. A distinction is made between primary control reserve, secondary control reserve and the minutes reserve [13].

In the upcoming transformation of electrical power production in Germany and Europe, wind and solar power have also gained importance. The main reason for this can be said to be a largely absent marginal cost of production because wind and solar power have no fuel costs at all. On the other hand, these two types of renewable energies are produced erratically and not always in line with demand patterns. For this very reason a range of flexibility options are required, for example: demand-side management, network expansion, energy storage and flexible power plants [14]. From a short-term perspective, flexible power could be met from fossil and renewable sources, but from a long-term perspective this task should be transferred exclusively to renewable power plants.

2.2.4 *Options for Integrating Biogas into the Future Power Supply*

Biogas plants are technically predestined for this challenge, because they provide a storable energy carrier (biogas), converted into engines showing a short response time for changing demand. Furthermore, the ability to serve control reserve as well as other system services makes them a versatile and rapidly realizable component of the future energy system [15].

Fundamentally, a flexible operating biogas plant is based on balancing the installed capacities of biogas production and biogas conversion into electricity, by converting a continuous energy provision from fermentation to an alternating mode of operation of the combined heat and power (CHP) provision with a higher amplitude in a shorter period of time (Fig. 2.3).

Depending on the demand of the energy market and marketing there are different options for this flexibilization from several minutes to seasonal shifts, requiring different adjustments to the concepts of the biogas plant (Table 2.3). In terms of marketing it has to be taken into account that biogas plants are usually too small to participate on their own in this marketplace, but specialized marketers have started to pool several biogas plants to contribute to the spot market. Additionally, for control reserve, several requirements have to be fulfilled to meet the pre-qualification for participating in this market. Because of the minimal size of bids (also for control reserve), biogas plants are pooled together to reach a minimum size, by marketers. For the moment, the main reasonable products to serve are negative secondary control reserve and both positive and negative minute reserves [16].

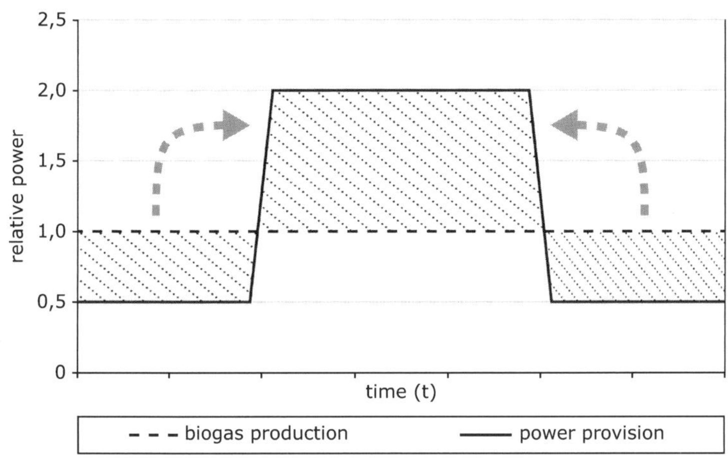

Fig. 2.3 Flexible mode of operation of biogas plants

Table 2.3 Different kinds of flexible power for biogas

Provision/Shift	Marketing	To balance	Additional technical demands
Up to 5 min	Secondary control reserve	Net frequency	Control gateway
5–15 min	Minute reserve	Net frequency	Control gateway
15 min – 6 h	Intraday	Forecast error	Gas storage
6–24 h	Day ahead	Residual load	CHP-capacity, heat storage
1–7 days	Day ahead	Macro weather situation	Feeding management
7–90 days	Day ahead	Seasonal demand	Feeding management

For the conception of flexible power generation from biogas different technical options are available and under development (discussed in Chap. 5).

2.2.5 Options for Integrating Solid and Liquid Biofuels into the Future Power Supply

Liquid biofuels could also be used in stationary engines, producing heat and power (CHP). In Germany, more than 2,000 CHPs using vegetable oils were installed in support of Renewable Energy Resource Act until 2012 (renewable energy law). Currently, approx. half of them are still operating with vegetable oils. Due to increasing prices, a large number of plants are not operating or have been converted to alternative renewables or fossil fuels [9].

CHPs are adapted for the flexible provision of heat (e.g. residential buildings, schools, market-gardens) as well as the demand-driven provision of electricity and are able to operate with vegetable oil as well as other liquid bioenergy carriers such as used cooking oil, animal fat, pyrolysis oil, biodiesel and bioethanol or solid and gaseous bioenergy carriers. Liquid biofuels can be stored easily and converted in standard diesel engines. As a result, they can provide flexible power for a comparatively high number of applications as biogas can. Liquid biofuels however are regarded as one of the options for the transition of the transport sector and because of this possible feedstock competition, the provision of electricity might not play a major role in the near future [17].

The thermo-chemical conversion of solid biomass plants could also contribute to flexible power generation. At the moment there are a lot of uncertainties about the theoretical potential and the determining factors that influence the flexible power provision of solid biomass plants, so that their contribution is expected to be more for the mid-term or long-term power adjustment strategies. A more detailed description is given in Chap. 4.

At the moment support schemes are neither envisaged for liquid nor for thermo-chemical conversion for greater flexibility among German renewable energy resources [18].

2.3 Heat Market

2.3.1 The Political Framework

In line with the European Renewable Energy Directive, Germany's national renewable action plan aims to increase the renewable energy share for heating from 9.1 % in 2009 to 15.5 % in 2020 [2]. This aim is supported by a combination of different regulations, including increasing energy efficiency and building insulation [19], integrating renewables for heat supply in new buildings [20] and investment support for low emission bioenergy stoves and boilers at different scales and for the district heating infrastructure [21]. The development of heat provision from biomass does not depend so much on support schemes but on emission control regulation, especially for the small-scale sector. The framework background tightens the emission protection enactment, which stipulates minimum energy efficiency and the lowering of emission limits for carbon monoxide and particulate matter [22]. Therefore, it becomes more expensive to adhere to strict limits, resulting in additional technical and economic expenses.

2.3.2 The Market Situation

In 2012 Germany's heat generation was covered by 10 % (or the equivalent of 144 TWh) from renewables [23]. A major proportion of that was contributed by solid biomass, see Fig. 2.4, with liquid and gaseous fuels playing only a minor role. Heat provision from biomass is mostly used for residential heating applications for a temperature level range of between 60 °C and 110 °C. Statistically, heat generation for industrial purposes from biomass, well above 100 °C today has a negligible impact [24]. The common form of usage in households is for small to medium stoves or boilers within the range of a few kilowatts to 100 kW. Heat generation on a larger scale above 100 kW includes heat generation solely for process and general heating purposes, as well as combined heat and power generation for electricity and heat provision.

2.3.3 Development Trends

The development of heat provision from biomass is influenced by the improvement in building insulation, emission reduction targets for biomass stoves and boilers, market conditions for combined heat and power installations based on biomass,

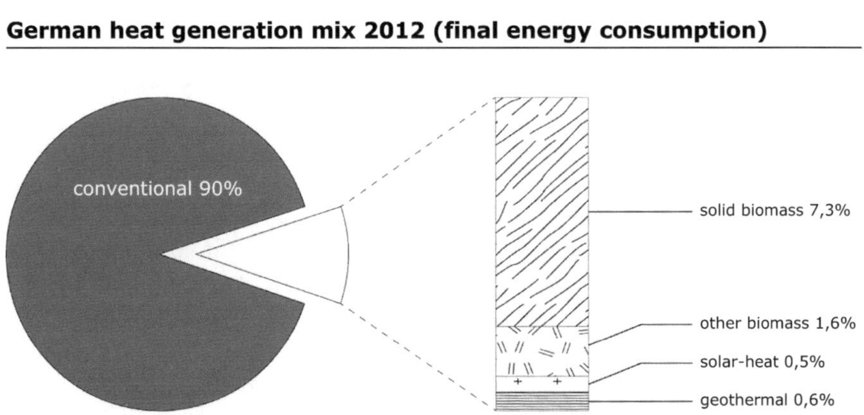

Fig. 2.4 Renewable energy provision for heating in Germany in 2012 (Data by [9])

targets for renewables in the heat sector and the market development of renewable heat provision from other sources (heat pumps, solar collectors etc.):

For the mid-term, it is presumed that a 45 % decrease for all heating sectors can be reached if ambitious restoration targets can be realized [25]. Most savings can be achieved in the building sector. For a current inventory of buildings it is supposed that most of the heat is needed for space heating and that the hot water supply is of minor relevance. If in the future, restoration and thereby insulation levels are improved, the demand for heating will generally decrease. This would also lead to a lower seasonal demand trend, with a more constant need shown for the amount of hot water usage. Furthermore, mixes with other renewable heat sources such as solar heat, geothermal heat, heat pumps and the conversion of excess electrical energy from power to heat will all increase [26]. Experiences from the past lead to the expectation that the transition in the heat sector will take place much slower than in the electricity sector. Today, the relative primary energy demand in Germany has an average of 260 kwh/(m^2*a), which is in line with the energy efficiency guidelines for new buildings (according to the EnEV 2009) that stipulates the use of less than 80–90 kWh/(m^2*a) [27].

2.3.4 Flexibility

The heat demand in the residential sector is characterised by seasonal, weekly and daily shifts. In case of a sole heat generation, the flexible heat provision covering exactly this demand is already state of the art. When compared with electricity, the storage and distribution is easy using hot water tanks, so called buffer-storage systems. Nevertheless, the specific demands and costs increase with decreasing heat provision capacities, which are expected in the future. In the case of a combined heat and power generation from biomass, it is usually the plant size and operational mode that are adapted to the demand characteristics of the heat sink. If CHP is going to change to a flexible power-guided operation mode, then it can be expected that there will sometimes be a mismatch between heat and electrical demand patterns. In this case, two different approaches are available for these conflicting goals. The first option is to add a secondary heat source to the given application, bridging emerging gaps in heat provision. This is state of the art in many concepts, where biomass serves the basic thermal load and already a pool of stoves or boilers act together to provide an all-season supply with variable demand. The second option is to install heat storage to counterbalance temporary mismatches. Both options require an individual strategy to quantify the particular need of additional demand and storage capacity respectively. A third way could be to create an integrated management that automatically takes into account both electricity and heat demand patterns [28].

2.4 Biofuels Market (Transport Sector)

Biofuels in the transport sector are characterized by a wide range of technical options with different maturity and market implementation stages. With regard to the actual market, biodiesel (fatty acid methyl esters – FAME), bioethanol and hydrogenated vegetable oils (HVO) have been introduced onto the market and will be described in the following. Additionally, biomethane can be used as a transport fuel, which is covered in Sect. 2.5.

2.4.1 The Political Framework

Numerous countries have defined national targets and mandates to increase the use of biofuels over recent years. The motivation of governments for setting such targets and implementing mandates consists mainly of (i) the desire for less dependence on importing fossil fuels, (ii) the security of supply and national added value and (iii) climate protection. The priority of these goals differs between countries and regions.

The European Union in particular has pursued an ambitious biofuel policy, adopting the relevant directives in 2009 to ensure a renewable fuels quota in the transport sector of 10 % (energy content) until 2020 in all member states [29]. In addition, specific sustainability criteria for biofuels have been defined. Biofuels and bio liquids taken into account for the quota should meet the specific requirements for the cultivation area and the cultivation practise for energy crops and a minimum of greenhouse gas emissions savings from the overall process chain (also see Sect. 3.3.2).

More sustainability requirements are discussed, for example social standards and emission factors to take into account the effects of indirect land use change.

There is also the possibility of double counting fuels from residues and waste materials [29]. The status of implementing the European Directive in national law differs between Member States. Therefore, the full impact has not yet been achieved [30].

Currently in Germany biofuels are to substitute 6.25 % (energy content) of the fossil diesel and petrol in the transport sector. In 2015 this regulation will be replaced by a quota for greenhouse gas reductions, which requires a 3 % reduction in greenhouse gas emissions from total fuel consumption using biofuels, 4.5 % from 2017 and 7 % from 2020 [1].

Different national and international standards define the minimum quality of current biofuels like e.g. biodiesel (FAME/HVO) and fuel ethanol or pure vegetable oil, bio-methane and Dimethylether. Thereby, a blending of fossil fuels with biofuels varies greatly depending on the market, e.g. in Brazil there are 18–25 % blends of ethanol in petrol [31] and a rapidly rising share in flex fuel vehicles (a tenfold increase in sales over the last 10 years, more than three million in 2013 [32]).

2.4.2 *The Market Situation*

The biofuel sector grew strongly over the period 2000–2010 –not only worldwide but also in Germany. Since 2010 output has increased moderately. Furthermore, a huge part of installed production capacity remains unexploited. So far, fossil fuels are largely substituted by bioethanol and biodiesel (Figs. 2.1, 2.2, 2.3, and 2.4). America is the focus of fuel ethanol production (in the U.S. primarily from corn and in Brazil primarily from sugar cane). In 2012 5 % of global production was realized in the European Union (mainly from cereals and sugar beet). Biofuels are globally traded, with imports of 1–2.5 million tons FAME and some 100,000 t to Europe over the last 3 years [32]. The global raw material base for biodiesel was composed of 28 % rapeseed oil, 32 % soybean oil, 22 % palm oil and 13 % animal fat and used cooking oil.

In Germany, pure vegetable oil and pure FAME (B100) represented a large proportion for energy consumption in the transport sector until 2008. From 2009 this decreased rapidly due to modified tax regulations and rising international prices for vegetable oils that were particularly used for trucks and agricultural machinery. In 2012 5.4 % (of energy content) of the transport fuels were substituted by 83.4 PJ/y biodiesel and 33.9 PJ/y bioethanol [33]. The domestic production included 101.1 PJ/y FAME and 16.8 PJ bioethanol [34]. Smaller amounts of biomethane were also used (see Sect. 2.5).

In 2011 approximately 3 % of the global transport energy was provided by biofuels. Their production volumes and the overall demand of the transport sector in 2011 are summarized in Figs. 2.5 and 2.6.

Current research and development activities focus on second generation biofuels from waste and residues and lignocellulosic biomass. At the same time, bioethanol and biodiesel consumed as a blend in road and rail transport has increased. The utilization of alternative fuels in further transport sectors such as shipping or aviation are increasingly under discussion [10].

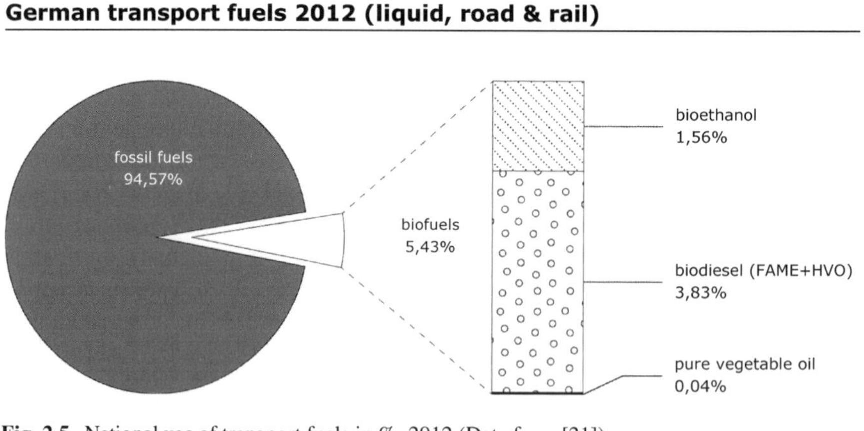

Fig. 2.5 National use of transport fuels in %, 2012 (Data from [21])

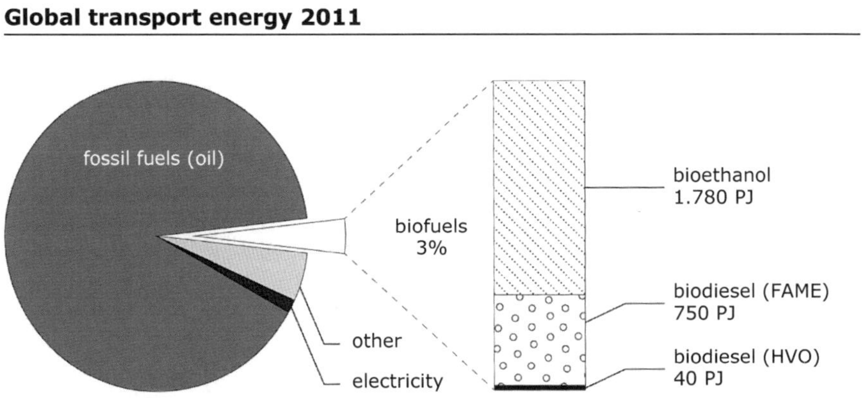

Fig. 2.6 Global energy demand of the transport sector, 2011 [6, 35]

2.4.3 Development Trends

The changeover of the biofuels quota in 2015 in Germany is associated with a number of uncertainties. On the one hand, the sub-mandates for biodiesel and ethanol fuel no longer exist. On the other hand, costs for reducing greenhouse gas emissions instead of purely the biofuel costs themselves will become important for shaping the market. Additionally, the publication of the proposal amending the renewable energy directive and the fuel quality directive in 2012 [36] as well as the policy framework for climate and energy [37] have caused great uncertainty regarding the political framework for biofuels after 2020.

By contrast, the transport sector in particular depends on propulsion systems that rely on carbon energy sources. Therefore, biofuels have a medium-term significant relevance in the transport sector, as well as other alternative fuels. Resource availability is of key importance for the specific alternatives. For example the use of biojet fuels in commercial aviation has received considerable attention in recent years. Consequently, almost all major commercial airlines and also some military sectors (i.e. in the USA), are heavily involved in testing and developing biojet fuels. Given the nature of the high quality drop-in fuels required in aviation, conversion technologies for the provision of jetfuels from biomass are rather limited, but not the main obstacle. Major advances are also necessary in terms of logistics, regulatory frameworks, quality assurance and the adoption of appropriate sustainability certifications, because any kind of biojet fuel market will become a global one [38].

2.4.4 Flexibility

Biofuels are suitable for storage and transportation over long distances in a similar way to fossil fuels. Biofuels and the majority of their feedstocks are traded on international markets as renewable energy sources as well as commodities (ethyl alcohol).

Thereby, the provision of biofuels is not coupled with utilization regarding the spatial and temporal scale.

The flexibility of biofuel production in terms of space and time is not as relevant as the flexibility of providing heat and electricity. The production of biofuels implies the conversion of biogenic raw materials to biogenic energy carriers generated by different technologies. Normally the raw materials, intermediate products and completed biofuels are suitable enough for storage and transport and therefore these commodities can be traded internationally. The production of biofuels is to a large extent decoupled from their utilization; thereby a systemic flexibility occurs regardless of the conversion technology.

Nevertheless, more flexibility will be more important within the production process with regard to the use of various raw materials in multi-feedstock plants. For the mid-term perspective, the coupled production of various products in one plant/bio-refinery will also be an interesting option for developing bioeconomy approaches. For certain resources commitment has already been shown. These circumstances enable an optimized operation of the plants regarding adjusted input and output depending upon the availability and the prices of raw materials as well as the demand and the revenue from products. Chapter 7 deals with technical options and requirements for a flexible production of liquid and gaseous biofuels.

2.5 Biomethane Market

Biomethane is defined as methane produced from biomass [35], with properties close to natural gas. When produced by thermal conversion (e.g. gasification), the methane-rich product gas is normally referred to as biobased synthetic natural gas (bio-SNG), whereas when it is produced by biological processes, including landfills, the initial product is raw biogas which must be cleaned (normally called upgrading) to reach the high methane content that is referred to as biomethane from biogas upgrading. Both processes can produce up to 99.9 CH_4 rich gases. Currently, the biochemical process is common practice, whereas thermochemically produced gas is still at the research and development stage. Focusing on the market aspects, this chapter will concentrate on biomethane produced by the biochemical process. Section 8.3 gives a more profound insight into the two different possibilities for producing biomethane from biochemical and thermochemical conversion.

2.5.1 The Political Framework

So far there is still no consistent biomethane strategy, certification and technical minimum standard in the European Union. Therefore, the situation regarding the biomethane market with its boundary conditions distinguishes between the member states, but it is still at a very low level – with Germany as the largest producer in Europe.

The German government has set the target of annually injecting 6 billion m^3 of biomethane into the natural gas grid by 2020, or 10 billion m^3 by 2030 respectively [39]. It is promoted for two markets: on the one hand biomethane is used for electricity, incited by the Renewable Energy Source Act (EEG), and on the other hand the fuel market, incited by the Biofuel Quota Act (BioKraftQuG).

Technical standards for biomethane ensure that the physical properties of the natural gas are met when it is injected into the grid. Primarily, they depend on the in-situ properties of the locally-used natural gas that the biomethane will substitute. In Germany, natural gas is distributed in two different gas qualities, low-gas (L-gas) and high-gas (H-gas) as a result of the different natural gas origins (mainly the North Sea and Russia). The specific national directions determine the gas standards for the upgraded biogas for injection into the gas grid. When used as a fuel, the DIN-51624 (automotive fuels – compressed natural gas) is applied. To pave the way for a European trade and exchange of biomethane, uniform product standards are being discussed intensively [40].

2.5.2 *The Market Situation*

In 2012 the consumption of natural gas in Germany amounted to 3.3 million TJ (equivalent to ~90 billion m^3 of natural gas), which is an increase in 4 % compared to 2011. Around 90 % of the natural gas used in Germany is imported from Russia, Norway and the Netherlands, see Fig. 2.7. With 21.5 %, natural gas covers more than 1/5 of the total energy consumption in Germany. Although the total energy

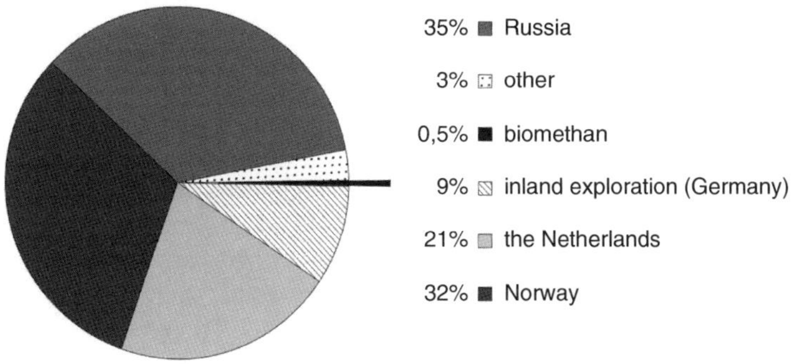

Fig. 2.7 The origin of natural gas in Germany [41, 42]

consumption has been declining since 1990, a slight increase in natural gas consumption during this time can be observed. Hence, the relevance of natural gas as an energy provider is also increasing. Natural gas is used for heating, electricity production and transport in the private, industrial and public sector. Furthermore it can be used in the chemical industry for other purposes. Because it is used for heat, the consumption of natural gas fluctuates depending on the seasons and the weather. Therefore, big gas storages, mostly underground are used [41, 43, 44].

In Germany the first biogas plants that were upgraded to biomethane were implemented in 2006. Since then a steady increase in these plants can be observed. At the end of 2012 120 biogas plants for biogas upgrading were in operation [9]. The majority of these plants injected the gas produced into the national gas grid and only a small percentage of plants directly delivered the biomethane to gas filling stations. About 413 million m^3 STP (standard temperature and pressure) of biomethane were injected in 2012 [42]. This accounts for approx. 0.5 % of the annual natural gas consumption in Germany.

2.5.3 Development Trends

With a further transition of the German energy system towards more renewables intended, an increasing importance of natural gas and a further development of the infrastructure are given. So far the development of biomethane injection over recent years has been slower than the pace required meeting the targets for 2020 and 2030. Some additional plant capacities for biomethane production and injection are in the planning and construction phases [45], but the supporting schemes for use in the electricity and fuel markets are currently under amendment and therefore uncertain [46].

Bio-SNG, as an alternative method for producing renewable methane from biomethane, is still in the development stage, see Sect. 8.3. Despite several research projects for an optimized production, so far no commercial plants are in operation.

2.5.4 Flexibility

As a result of the chemical and physical properties of natural gas and biomethane as well as their similarity, numerous flexibility properties are feasible. One major advantage is that biomethane benefits from natural gas storage in the gas grid itself as well as in several underground storages [44]. Therefore, it is eminently suitable, because of sufficient experience of natural gas, for peak and demand-driven loads for heat and electricity provision. However, up to now this potential has not been used.

2.6 Cross-Sectoral Markets from Power to Heat, to Gas and to Liquid

With growing shares of renewable energies, in the future the individual energy sectors (electricity, mobility, heating) will merge more and more together. An initial point would be the electricity sector, because of its universal characteristic to make every known kind of final energy out of it. Two example forces are the introduction of e-mobility and the promotion of heat pumps. Another one is that with increasing shares of photovoltaic and wind power, it is foreseen that there will be periods with an excessive power feed-in in relation to the actual demand.

In line with the projected expansion of fluctuating energy sources for the year 2032, it is expected that there will be an amount of excess energy between 2 % and 18 % of annual solar and wind production. Great uncertainties are still caused by technical incentives for the realisation of flexibility options for fossil power plants and the transformation of biomass plants from base load operation to alternating production [47]. The current situation is characterized by a moderate but growing amount of excess energy caused by network congestions (Table 2.4).

To make such overshoots available for the energy supply in the future, the conversion of power into heat, gas and liquids is being intensively discussed and tested in Germany.

The first option which is already at the point of market entry is the technology of so-called power to heat (P2H), which is simply a way of producing heat from excess energy [48]. For example, biogas plants which serve negative control power could use this option to serve negative loads while keeping their thermal output for connected heat sinks. In this case, the P2H is dimensioned at half of the installed capacity of the power provision unit, so that if there is a request for negative power, the plant can switch down to half load and the produced electrical power is directly and within the P2H converted into heat. Consequently, the plant can then serve a full power hub, by maintaining thermal generation and a faster reaction to recalls in both directions in case of a total switch off.

In the distant future, the so-called power to gas technology (P2G) could be suitable to transform excess power into gas [49]. In a first step, electrical energy is used for electrolysis to produce hydrogen. This hydrogen can be used as it is or could be further transformed into synthetic methane or liquid fuels like methanol (power to liquid – P2L) [50]. This technology is under development and could be an option in the future to substitute fossil fuels in mobility applications which have

Table 2.4 Excess-energy in the German power network (Data source: German network agency [16])

	2009	2010	2011	2012	2013
Excess energy [GWh]	74	127	421	385	555
Share of renewable feed-in [%]	0.10 %	0.16 %	0.41 %	0.29 %	0.44 %

a mandatory need for chemical energy storage, for example aeroplanes or heavy load transportation.

If the renewable gas from electrolysis is injected into the gas grid, the tolerable content of hydrogen in such grids will be limited for technical reasons, then it should be necessary to convert hydrogen, together with carbon dioxide into synthetic methane [51]. Biogas plants, especially those who separate raw biogas into biomethane and a CO2 rich offgas, can also provide a renewable carbon source (see also Sect. 2.4.).

2.7 Conclusion

Bioenergy is a relevant and well established energy carrier for power and heat provision and as a substitute for fossil transport fuels. With regard to their development towards smart flexible systems, the demand of the different markets is different in terms of both quality and dynamics:

The German power market is in a dynamic transition towards renewables and already needs flexible power to balance the volatile wind and solar power and stabilise the power grid. On the other hand, relevant renewable energy installations in the form of biogas plants and power generation plants from solid biofuels might build the basis for a flexible power generation based on biomass. For biogas market incentives have also been established in the form of the Renewable Energy Sources Act in 2012, with the flexible premium to provide incentives for investments in flexible power generation. The flexibilisation of power provision from biomass can therefore be seen as an interesting and promising short-term option for this transition.

By contrast, the transition of the heat market towards renewable energies has been much slower and the characteristics for the future demands on bioenergy in this sector are not that well defined yet. Improved insulation is expected, combined with a decreasing specific heat demand on the one hand and an integration of additional renewable heat supply units on the other. In terms of long term development, the future heat provision from biomass might be faced with smaller conversion units and additional flexibility.

For the substitution of fossil fuels in the transport sector and for natural gas applications, fuel provision does not depend on varying frame conditions demanded by a flexible provision. Here, the challenge for future demand is more in the field of sustainable resource availability and the stepwise production and implementation of different products for matter and energy uses in biorefineries. Moreover, biomethane has not yet been fully implemented on the market.

In the future, the sectoral analysis of power, heat, transport and gas markets will only deliver half of the picture because all market segments are expected to merge. As a result, some of the flexibility needs can be shifted between the different sectors. Especially for example of the upper excess electrical energy can be converted into thermal or chemical energy and meet some of the demand for heat or fuel consumption. Biomass, especially biogas, can link the sectors by providing the renewable carbon source for the provision of renewable gases as chemical energy storages. It can also provide the option to balance long-term fluctuations in power production, and seasonal storage functionality, which could not be covered by conventional flexibility options or common storage technology. From today's perspective this can be regarded as a second step of the transition, based on flexible technologies and concepts, which are described in the following chapters.

References

1. BImSchG, BImSchG – Bundes-Immissionsschutzgesetz in der Fassung der Bekanntmachung vom 26 Sept 2002 (BGBl. I S. 3830), zuletzt geändert am 3 Nov 2011
2. NREAP Germany, National renewable energy action plan in accordance with directive 2009/28/EC on the promotion of the use of energy from renewable sources (Berlin, 2010)
3. AGEB, *Auswertungstabellen zur Energiebilanz für die Bundesrepublik Deutschland 1990–2012* (Arbeitsgemeinschaft Energiebilanzen e.V, Berlin, 2013)
4. Federal Republic of Germany, National renewable energy action plan in accordinace with directive 2009/28/EC on the promotion of the use of energy from renewable sources (Berlin, 2010)
5. EU, Directive 2009/28/EC of the European Parliament and of the council – 2009/28/EC (2009). The European Parliament and the council of the European Union
6. EEG 2009 — Gesetz für den Vorrang Erneuerbarer Energien (2012)
7. N. Szarka, F. Scholwin, M. Trommler, J.H. Fabian, M. Eichhorn, A. Ortwein, D. Thrän, A novel role for bioenergy: a flexible, demand-oriented power supply. Energy **61**, 18–26 (2013)
8. Agency for Renewable Energies, www.unendlich-viel-energie.de, chart, 2013, accessed 2014
9. J. Daniel-Gromke, V. Denysenko, P. Sauter, K. Naumann, M. Scheftelowitz, A. Krautz, M. Beil, W. Beyrich, W. Peters et al., *Stromerzeugung aus Biomasse* (BMU, Leipzig/Berlin/Halle/Kassel, 2013)
10. K. Naumann, K. Oehmichen, M. Zeymer, *Monitoring Biokraftstoffsektor, DBFZ Report* (Nr. 11) (Deutsches Biomasseforschungszentrum, Leipzig, 2014)
11. EEX, *Power Terminmarkt*. https://www.eex.com/de/produkte/strom/power-terminmarkt. abgerufen am 11 Mar 2014
12. EEX, *Power Spot Market*. https://www.eex.com/de/produkte/strom/power-spot-market. abgerufen am 11 Mar 2014
13. Regelenergie.net, *Allgemeines zur Regelleistung – Technische Aspekte*. https://www.regelleistung.net/ip/action/static/techaspects. abgerufen am 11 Mar 2014
14. U. Leprich, E. Hauser, K. Grashof, L. Grote, M. Luxenburger, M. Sabatier, A. Zipp, *Kompassstudie Marktdesign Leitideen für ein Design eines Stromsystems mit hohem Anteil fluktuierender Erneuerbarer Energien* (2012)
15. R. Mackensen, K. Rohrig, H. Emanuel, *Das regenerative Kombikraftwerk* (ISET, Kassel, 2008)
16. S. Bofinger, M. Braun, C. Costa Gomez, J. Daniel-Gromke, N. Gerhardt, K. Hartmann, D. Kirchner, T. Reimann, Y.-M. Saint-Drenain; u. a., *Die Rolle des Stromes aus Biogas in zukünftigen Energieversorgungsstrukturen* (Fraunhofer-Institut für Windenergie und Energiesystemtechnik IWES, Hanau, 2010)
17. D. Thrän, O. Arendt, J. Ponitka, J. Braun, M. Millinger, V. Wolf, M. Banse, R. Schaldach, J. Schüngel, S. Gärtner, N. Rettenmaier, K. Hünecke, K. Hennenberg, B. Wern, F. Baur, U. Fritsche, H.-W. Gress, Meilensteine 2030 – Elemente und Meilensteine für die Entwicklnug einer tragfähigen und nachhaltigen Bioenergie, Energetische Biomassenutzung No. 18, 2014
18. EEG 2012 — Gesetz für den Vorrang Erneuerbarer Energien (Erneuerbare-Energien-Gesetz – EEG)Konsolidierte (unverbindliche) Fassung des Gesetzestextes in der ab 1. Januar 2012 geltenden Fassung*

19. EnEG, Gesetz zur Einsparung von Energie in Gebäuden (Energieeinsparungsgesetz – EnEG, 2013)
20. EEWÄrmeG, Gesetz zur Förderung Erneuerbarer Energien im Wärmebereich (Erneuerbare-Energien- Wärmegesetz – EEWärmeG, 2011)
21. BAFA, *Heizen mit erneuerbaren Energien – Marktanreizprogramm*. http://www.bafa.de/bafa/de/energie/erneuerbare_energien/. abgerufen am 17 June 2014
22. I. Hartmann, V. Lenz, M. Schenker, C. Thiel, M. Kraus, M. Matthes, U. Roland, R. Bindig, W.-D. Einicke, *Katalytisch unterstützte Minderung von Emissionen aus Biomasse-Kleinfeuerungsanlagen, DBFZ Report Nr. 6* (DBFZ Deutsches Biomasseforschungszentrum gemeinnützige GmbH, Leipzig, 2011)
23. D. Böhme, T. Nieder, T. Rüther, P. Bickel, *Entwicklung der erneuerbaren Energien in Deutschland im Jahr 2012* (Bundesministerium für Umwelt, Naturschutz und Reaktorsicherheit (BMU), Berlin, 2013)
24. C. Viehmann, T. Westerkamp, A. Schwenker, M. Schenker, D. Thrän, V. Lenz, M. Ebert, DBFZ, *Ermittlung des Einsatzes biogener Festbrennstoffe im GHD-Sektor* (Unveröffentlichter Endbericht) (DBFZ Deutsches BiomasseForschungsZentrum gemeinnützige GmbH (DBFZ), Leipzig, 2011)
25. J. Nitsch, T. Pregger, T. Naegler, D. Heide, D.L. de Tena, F. Trieb, Y. Scholz, K. Nienhaus, N. Gerhardt; u. a.: *Langfristszenarien und Strategien für den Ausbau der erneuerbaren Energien in Deutschland bei Berücksichtigung der Entwicklung in Europa und global*. http://www.erneuerbare-energien.de/fileadmin/ee-import/files/pdfs/allgemein/application/pdf/leitstudie2011_bf.pdf. abgerufen am 24 Apr 2013
26. W. Schulz, C. Brandstätt, A. Hagemeister, T. Holzfuss, J. Gabriel, *Flexibilitätsreserven aus dem Wärmemarkt* (Fraunhofer-Instituts für Fertigungstechnik und Angewandte Materialforschung IFAM, Bremen, 2013)
27. U. Bigalke, H. Discher, H. Lukas, Y. Zeng, K. Bensamnn, C. Stolte, *dena-Gebäudereport 2012* (dena, Berlin, 2012)
28. E.-M. Klotz, M. Koepp, G. Steudle, *Maßnahmen zur nachhaltigen Integration von Systemen zur gekoppelten Strom- und Wärmebereitstellung in das neue Energieversorgungssystem* (Prognos AG, Berlin, 2013)
29. European Parliament; European Council (Hrsg.), Directive 2009/28/EC of the European Parliament and of the Council of 23 April 2009 on the promotion of the use of energy from renewable sources and amending and subsequently repealing Directives 2001/77/EC and 2003/30/EC (2009)
30. L. Pelkmans, N. Devriendt, L. Goovaerts, P-P. Schouwenberg, *Prospective Study: Implementation of Sustainability Requirements for Biofuels and Bioenergy and Related Issues for Markets and Trade* (Final Report) (IEA Bioenergy, Paris, 2012)
31. S. Barros, *Brazil Biofuels Annual Report 2013, GAIN Report* (Nr. BR13005) (USDA Foreign Agricultural Service, Sao Paulo, 2013)
32. F.O. Licht, World Ethanol & Biofuels Report Bd. 2008–2013 (Informa plc, London)
33. BAFA, *Amtliche Mineralöldaten* (Bundesamt für Wirtschaft und Ausfuhrkontrolle, Eschweiler, 2013)
34. Destatis, *Wirtschaftsbereiche – Energie – Erzeugung – Statistisches Bundesamt*. https://www.destatis.de/DE/ZahlenFakten/Wirtschaftsbereiche/Energie/Erzeugung/Tabellen/Biotreibstoffe.html. abgerufen am 3 Sept 2012
35. ISO 16559:2014: Solid biofuels – Terminology, definitions and descriptions (Abstract ISO 16559:2014)
36. European Commission, Proposal for a Directive of the European Parliament and of the Council amending Directive 98/70/EC relating to the quality of petrol and diesel fuels and amending Directive 2009/28/EC on the promotion of the use of energy from renewable sources (2012)
37. European Commission, A policy framework for climate and energy in the period from 2020 to 2030 (2014)
38. M. Seiffert, D. Thrän, *The Potential and Role of Biofuels in Commercial Air Transport-Biojetfuel* (IEA Bioenergy, London, 2012)

39. GasNZV, *Gasnetzzugangsverordnung (Gas Network Access Ordinance)* (Verordnung über den Zugang zu Gasversorgungsnetzen, 2012)
40. DENA, press release: Biogas registers combine to form network – International cooperation to drive biomethane trade forward (2013)
41. BAFA, *Entwicklung der Erdgaseinfuhr in die Bundesrepublik Deutschland – Bilanzen 1998–2012* (Bundesamt für Wirtschaft und Ausfuhrkontrolle, Eschborn, 2013)
42. BNetzA, *Biogas-Monitoringbericht 2014* (Bundesnetzagentur, Berlin, 2014) http://www.bundesnetzagentur.de/cln_1411/DE/Sachgebiete/ElektrizitaetundGas/Unternehmen_Institutionen/ErneuerbareEnergien/Biogas/Biogasmonitoring/biogasmonitoring-node.html;jsessionid=113ACF9D35D31A4367B2E9771FE3496E
43. AG Energiebilanzen, *Auswertungstabellen zur Energiebilanz für die Bundesrepublik Deutschland 1990 bis 2012* (Arbeitsgemeinschaft Energiebilanzen e.V., Berlin, 2013)
44. EEK, *Underground Gas Storage in Germany, Erdöl, Erdgas, Kohle (EEK)* (Urban-Verlag AG, Hamburg, 2013)
45. DENA, *Branchenbarometer Biomethan – Daten, Fakten und Trends zur Biogaseinspeisung* (Deutsche Energie-Agentur, Berlin, 2013)
46. EEG, Entwurf eines Gesetzes zur grundlegenden Reform des Erneuerbare-Energien-Gesetzes und zur Änderung weiterer Bestimmungen des Energiewirtschaftsrechts – Draft Law, 2014
47. W-P. Schill, *Integration von Wind- und Solarenergie: Flexibles Stromsystem verringert Überschüsse, DIW Wochenbericht* (Nr. 34) (DIW, Berlin, 2013)
48. H-M. Groscurth, S. Bode, "Power-to-heat" oder "Power-to-gas"?, arrhenius Institut für Energie- und Klimapolitik (2013)
49. M. Jentsch, T. Trost, M. Sterner, Optimal use of power-to-gas energy storage systems in an 85% renewable energy scenario. Energy Proc. **46**, 254–261 (2014), 8th International Renewable Energy Storage Conference and Exhibition (IRES 2013)
50. R. Meyer-Pittroff, *Power-to-Liquid (Methanol) zur Energiespeicherung* (Technische Universität München, München, 2013)
51. N. Grimm, I. Zoch, U. Weiß, *Integration erneuerbaren Stroms in das Erdgasnetz* (dena, Berlin, 2012)

Chapter 3
Biomass Resources and Sustainability Issues for a Flexible Bioenergy Provision

Stefan Majer and Daniela Thrän

Abstract Biomass available for the flexible provision of bioenergy is a major factor in discussing the potential contribution flexible bioenergy systems could make to the overall energy system. Even though the quality of the biomass used has an impact on the potential availability of biomass, it might not be the most decisive factor. More important is the origin of the biomass since the production of biomass has a complex impact on land and land use and can also provoke change in land use. Many studies have been carried out to estimate future biomass potentials. Their results differ greatly, due to different methods, definitions and assumptions regarding the scope of the studies. Sustainable provision of biomass is a precondition for smart bioenergy supply. With liquid biofuels as a starting point, a number of certification schemes have been developed over recent years and recognised by the European Commission. The future development of these schemes and possible expansion to the whole agricultural or forestry sector will also influence the future biomass potentials of energy crops. This underlines the uncertainty surrounding the future potential of energy crops. In regards to smart bioenergy provision, one possible option is to make (existing) larger production units using energy crops more flexible by widening their product portfolio. To satisfy the specific technical demands of flexible provision greater quantities of feedstock will be required.

S. Majer (✉)
Deutsches Biomasseforschungszentrum GmbH – DBFZ,
Torgauer Str. 116, 04347 Leipzig, Germany
e-mail: stefan.majer@dbfz.de

D. Thrän
Department of Bioenergy, Helmholtz Centre for Environmental Research – UFZ,
Permoset Straße 15, 04318 Leipzig, Germany

Deutsches Biomasseforschungszentrum – DBFZ, Torgauer Straße 116,
04347 Leipzig, Germany

Bioenergy Systems, University of Leipzig, Grimmaische Straße 12,
04109 Leipzig, Germany
e-mail: daniela.thraen@ufz.de

D. Thrän (ed.), *Smart Bioenergy*, DOI 10.1007/978-3-319-16193-8_3

3.1 Introduction

The resources available for the flexible provision of bioenergy are a major factor in discussing the potential contribution flexible bioenergy systems could make to the overall energy system. In general, bioenergy production is based on a wide variety of technologies and feedstock:

- For thermochemical conversion a dry, carbon rich source is needed. This includes woody biomass and lignocellulosic material, taken from forestry or wood processing industry, straw and husks from agricultural production and different residues from gardening, land scape management etc.
- For biochemical conversion biomass with high water and sugar and/or starch content (depending on the fermentation process) is favourable. Typical feedstock for biochemical conversion include residues from livestock production and food processing industry, organic waste and different energy crops, including sugar cane, sugar beet, maize and grain crops.
- For physico-chemical conversion a source with high oil content, such as palm, rape seed, sunflower seed etc. Additionally, there are small quantities of used cooking oil available today. Algae feedstock is also discussed as a long term source solution.

Even though the quality of the biomass used has an impact on the potential availability of biomass, it might not be the most decisive factor. More important is its origin as the production of biomass has a complex impact on land, land use and can provoke change in land use.

The discussion surrounding biomass potentials for this specific sector of bioenergy production is therefore closely linked to the wider discussion on biomass potentials for bioenergy. In regards to this matter, important aspects to consider include the potential environmental impact of biomass production and the measures needed to avoid or minimise this impact. There are different expectations of the realistic future use of biomass, as a product of domestic production or import, in different countries [30]. Therefore, the assessment of biomass potentials has to take the global perspective into account. The objective of this chapter is therefore twofold: The first section will summarise the ranges of biomass available for the production of bioenergy and will describe the main drivers influencing the scenarios used for assessing biomass potentials. The second section of this chapter will touch upon the important topic of potential environmental impact both on a local and global scale. In the first part of this chapter, a number of potential local environmental aspects from the production of biomass in agricultural systems will be discussed. The environmental impact of biomass production on a global level is addressed within a number of sustainability certification schemes for biomass and bioenergy production. The second part of this chapter includes a discussion on recent development in this area as well as a brief summary of existing schemes and the environmental criteria included in their standards.

3.2 Biomass Potentials and Drivers

Biomass potentials for bioenergy production have been the subject of numerous studies considering a variety of geographical areas and resolutions, biomass assortments, assumptions and time frames. Available results therefore differ greatly, with some reports concluding that biomass has no potential while others conclude that biomass has huge potential and could satisfy the world energy demand multiple times offering a long term solution. Since each of the studies available considers different scoping questions and thus different framework conditions, the results of existing potential assessments are difficult to compare. Amongst other parameters, the individual definition of biomass potential is an important point for consideration and has a decisive effect on the outcome of the assessment of biomass potential. Different definitions of biomass potentials exist. Kaltschmitt et al. 2009 [18] distinguishes between:

- Theoretical potential describes the theoretically usable physical energy supply (e.g. all energy stored by phyto- and zoomass) of a given region in a certain time span. It is solely defined by the limits of physical use and thus represents the upper limit of biomass' theoretically feasible contribution to energy supply. Due to inseparable technical, ecological, structural and administrative barriers usually only a minimal realisation of its theoretical potential is possible.
- Technical potential is a function of the abilities of the technology which is currently available. Additionally, technical potential takes into account structural, environmental (e.g. nature conservation areas) and other non-technical restrictions. Technical potential therefore describes renewable energy's possible contribution to the satisfaction of energy demands for technical purposes, depending on time and location. As technical potential is primarily dependant on technical constraints it is less subject to fluctuations than the economic potential. The results summarized in this chapter represent technical potentials.
- Economic potential describes the proportion of the technical potential that is economically exploitable according to the given basic conditions. Since there are different ways to assess the economic efficiency of an option, there is always a multitude of economic potentials. Furthermore, continuously changing basic conditions (e.g. oil price changes, changing CO_2-tax models, energy and ecotaxes) influence economic potential.

The following figure summarizes the results from 19 different studies on the potential of biomass for bioenergy including energy crops, organic residues and waste. The majority of these publications focus on long-term energy potential of biomass (2050 and even 2100). Few publications specifically address the short and mid-term potential (2020 and 2030).

The figure shows that energy crop potentials are the most uncertain. Residue potentials are much less variable and range between 20 and 50 PJ/a [34]. Particularly from a long term perspective, the ranges for energy crop potentials are extremely wide (ranging between 0 EJ/year and values of 1,272 EJ/year for very optimistic assumptions). In comparison, the global primary energy supply in 2012 was approximately 500 EJ [14].

The definition of biomass in regards to the category forest residues is not consistent between the studies analysed. While some authors limit the use of this term only to residues obtained from thinning and logging, industrial production processes, and waste, others also include the annual forest increment. Due to the existing disagreements regarding the definition of forest derived residues, this resource shows the most significant changes across studies of a residue fraction. Its potential ranges from zero to 150 EJ/year in 2050. It should also be noted that the biomass potential discussed in recent publications tends to range between 50 and 200 EJ (cf. [3, 29]). The potential of biogenic wastes and residues is an important fraction for a sustainable supply of bioenergy through biomass. According to [34] this potential amounts to approx. 50 EJ/a. Given the intense debate questioning whether a large scale use of biomass to produce bioenergy is sustainable, these ranges seem to indicate greater potential for future bioenergy strategies and scenarios.

Besides methodical differences between the different studies, two additional points regarding the summarized results for the biomass residue potentials and the energy crop potentials have to be considered. Firstly, the studies analysed in Fig. 3.1 include different material flows (i.e. forest and agricultural residues, industrial residues, waste streams considering the demand of renewable materials for certain production processes, nutrition cycles, etc.) under the term biomass residues.

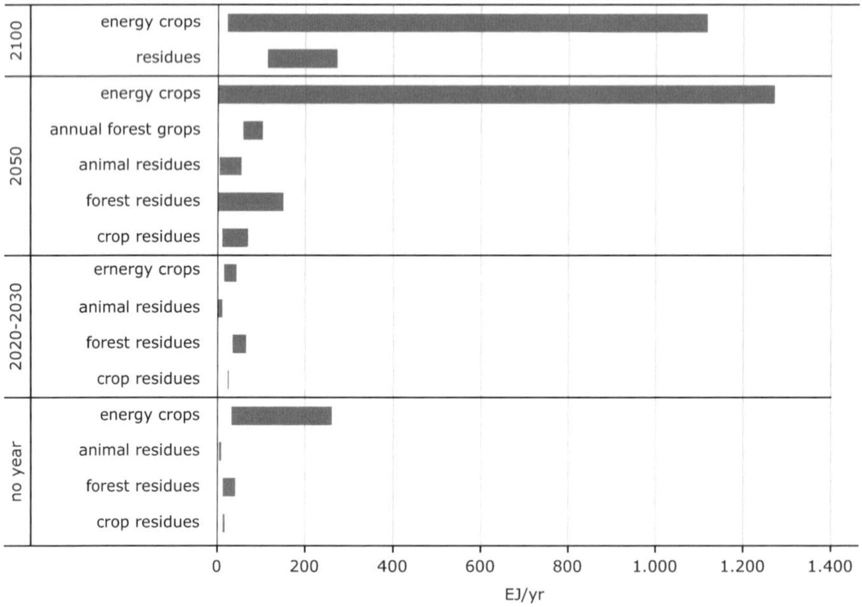

Fig. 3.1 Ranges of biomass potential of different resource fractions and years; from [31], data from [1, 2, 4, 7, 9, 11–13, 15, 17, 20, 21, 26–28, 35–37]

Secondly, the future demand of arable land for food production, one of the main parameters determining the potential of energy crops is considered differently by the studies considered in Fig. 3.1. On the one hand, an increasing population together with a change in food consumption patterns and an increasing urbanization lead to an additional demand for arable land for food production. On the other hand, increasing yields might reduce the specific area demand for the food production significantly. Besides the utilisation of biomass from residues, the production of energy crops in agricultural production systems is of high significance. The land availability for energy crop production depends on the overall amount of available agricultural land and the demand of land for food and fodder production. There are various drivers which influence the present and future food and feedstock demand. Their magnitude depends on the climate zone, the soil quality and specific local conditions. However, the main factors are universally valid in a global context. In the following table, an overview of the main influential factors is given (Table 3.1).

Out of these factors, the most important ones are the growth in (global) population, the future per-capita consumption – both driven by worldwide economic growth – and developments in the yield for food, fodder and biomass production. Climate change and its impact on agriculture production will also be an important factor which is however difficult to quantify. In order to estimate biomass potential these factors need to be considered altogether. The assessment of the future potential of biomass involves a great deal of uncertainty and therefore raises a complex question.

An important aspect in the general debate about the sustainability of biomass production for bioenergy (including flexible bioenergy provision) is the potential environmental impacts associated with its production. Besides global environmental

Table 3.1 Overview of major variables and drivers for biomass potentials

Variable	Explanation
Development of crop yields	High yields reduce the size of the agricultural area required for food and fodder cultivation, thus land and therefore energy crop potentials increase
Population growth	Determines the demand for foodstuffs and thus the area available for biomass cultivation
Development of livestock numbers	Influences the size of the area required for fodder cultivation
Impervious surfaces	Reduce the total available area of arable land
Per capita consumption	Extent of food consumption per capita influences the size of the area required for food and fodder cultivation
Foreign trade balance	Determines the level of self-sufficiency and thus the size of the available area
Conservation of land development	Determines the availability of land through changes in cultivation management
Climate change	May result in decreasing yields due to changing climatic conditions, amongst other things

aspects, such as greenhouse gas (GHG) emissions resulting from the intensification of agricultural processes or the effects of changes in land use, a number of local environmental aspects should be considered. The next subchapter will therefore focus on the discussion of a number of local environmental aspects of biomass production.

3.3 Environmental Aspects of Biomass Production and Certification

General Aspects

The concept of environmental sustainability is broadly defined per se. It includes, in simple terms, the preservation of nature and the environment for future generations. This objective affects a variety of aspects such as the conservation of biodiversity, climate protection, landscape maintenance, the protection of natural areas and the careful use of natural resources as well as the consideration of numerous additional environmental aspects. Furthermore, the close relationship and interdependency between all these environmental aspects makes the discussion about sustainable production of flexible bioenergy provision even more complex. Despite this complex concept of sustainability and the interdependencies described above, the political and social discourse is mainly focused on aspects such as climate protection or biodiversity. Major principles of sustainable biomass cultivation are important elements in various regulations on European and national level (e.g. in Germany). Since the production of energy crops to produce flexible bioenergy is part of general agricultural production systems in Europe, the respective regulatory framework for the agricultural sector and therefore many of the regulatory documents mentioned are relevant for the production of energy crops. The cross-compliance rules at European Union (EU) level for example include a number of requirements for good agricultural practice in biomass production. On a national level, additional measures (e.g. the Federal Immission Control Act[1]) supplement the European regulations by adding additional requirements (e.g. on thresholds for local emissions and environmental impacts). Compliance with the existing legal requirements enforcing sustainable cultivation of biomass is therefore a prerequisite for the establishment and running of energy plants to produce flexible bioenergy (e.g. in biogas systems).

One of the main drivers for the promotion of bioenergy systems in recent years has been the strong interest to reduce anthropogenic GHG-emissions. This aspect and potential benefit of bioenergy production in particular has been subject to intense debate in the recent past. In particular, the potential GHG emissions from

[1] Siebzehnte Verordnung zur Durchführung des Bundes-IMmissionsschutzgesetzes (Verordnung über die Verbrennung und Mitverbrennung von Abfällen – 17. BImSchV) vom 29.Januar 2009 (BGBl. I S.129)

the effects of change in land use (e.g. from the conversion of natural lands into cropland) as well as an inefficient use of biomass resources can reduce or even negate the potential GHG savings from the use of bioenergy (e.g., [8, 25]). This discussion illustrates the importance and need for the implementation of additional legal requirements for the biomass sector and bioenergy production. In regards to liquid biofuels, the Renewable Energy Directive has introduced a number of additional sustainability criteria [5]. The introduction of precise GHG-mitigation thresholds for biofuel systems are, among other criteria (e.g. requirements regarding good agricultural practice and the definition of no-go-areas) one key element of this directive. Expanding these criteria to all areas of bioenergy and biomass production (including the material use of biomass) in the years to come appears to be important and necessary in order to further increase the sustainability of biomass production.

3.3.1 Potential Environmental Issues Surrounding the Production of Energy Crops for the Provision of Flexible Bioenergy

The cultivation of energy crops for the production of flexible bioenergy follows the existing legal framework for agricultural production at European and national levels. Requirements and rules for good agricultural practice at national level (e.g. the Plant Protection Act (PflSchG),[2] the Federal Soil Protection Act (BBodSchG) and the Fertilisation Act (DüV) on a national level in Germany) are also relevant for the cultivation of energy crops (e.g. for biogas production) must also be adhered to for the cultivation of other crops (e.g. feed and fodder). From a legal perspective, the binding nature of these rules basically ensures the avoidance of severe negative effects on the soil from biomass cultivation for bioenergy production. This is also true for high potential emissions from the conversion of natural land into areas for agricultural production (land use change, LUC). In countries outside of the European Union this important aspect can be tackled with the help of sustainability certification schemes for biomass and bioenergy.

However, in addition to the general legal framework on the definition of good agricultural practice in the context of the existing cross-compliance, crop rotation systems, which include a wide variety of crops, show additional environmental benefits. Systems which include different shallow and deep rooting plants or plants providing carbon benefits etc. in particular can help to reduce potential risks from weeds, fungal diseases and other pests and to increase the overall nutrient and water availability in the soil compared to the monocultures. The use of elements such as plant protection agents might therefore be reduced. Furthermore, the cultivation of energy crops as part of a diverse crop rotation system can help to increase

[2] Gesetz zum Schutz der Kulturpflanzen (Pflanzenschutzgesetz in der Fassung vom 02. Dezember 2014).

soil cover and thus reduce soil erosion and the risk of nitrogen loss during the winter period [32, 33]. The potential effects of energy crop production on some of the environmental aspects discussed are explained in the next few paragraphs.

Soil Erosion

Improper or unsustainable cultivation of agricultural land can trigger and aggravate soil erosion through wind or water. The consequence of this development is the erosion of the fertile upper soil classes, a resulting degradation of the soil and possible soil devastation. These effects can influence the productivity of the soil and, due to the discharge of nutrients, also contribute to other undesired effects such as the eutrophication of water bodies. In addition to various other factors, the canopy in particular has a significant influence on the effects of possible soil erosion. For this reason, the integration of crop production for flexible bioenergy provision into diverse crop rotation systems and the avoidance of monocultures are of high importance.

Humus Balance

The term humus encompasses the whole of the soil organic matter, the designated organic residues and their degradation products. The maintenance or slight increase of the humus content on agricultural land due to a positive influence on the carbon and nitrogen turnover is one of the basic requirements and safeguards to ensure mid and long term soil quality [22].

Due to the permanent activity of soil organisms, the soil humus content is subject to a constant process of reduction, conversion and construction. This process is influenced by various additional parameters such as the type of vegetation, climatic factors or land use. The Humus content as well as its composition significantly influences the soil characteristics. The humus content in arable soils is therefore characterized, for example by intensive mixing with minerals and is approximately between 1.8 % and 2.5 %. The oversupply of soils with organic matter is just as detrimental as a lack of supply. This oversupply can result in uncontrolled mineralization and increased nutrient loss. For this reason, the overall humus balance of a cropping system is of significant importance. The supply of organic material (e.g. fermentation residues, green manure, straw, manure, slurry) can compensate possible humus deficits. It should be noted, that this basic principle of good agricultural practice should be applied and considered regardless of the final use of the produced agricultural goods (e.g. bioenergy production, food, fodder, industrial use). In contrast to other production systems, one advantage of the production of energy crops for biogas (e.g. for flexible energy provision) is the option of using the digestate (as co-product of the biogas process) to return a significant proportion of nutrients and carbon to the agricultural land [24].

Pesticides and Fertilizers

As described previously, the production of biomass for the provision of flexible bioenergy has to follow the very same existing governmental legal framework conditions as the production of other biomass for feed, fodder or other purposes. This includes requirements and thresholds for the use of pesticides or fertilizers. In some respects, systems for the production of feedstock for the supply of flexible bioenergy (e.g. silage maize) can be slightly modified in order to increase the methane yield of the energy crops. Some examples include the use of slightly higher seed densities, earlier harvest times at lower degrees of lignification and an ideal dry matter content as well as reduced chop lengths (to improve the enzymatic degradation of the biomass during fermentation) [10].

A decisive factor for the biomass and thus biogas yield is fertilizer management. Compared to conventional cropping systems (e.g. the production of wheat for food production) slight adjustments to the total amount of fertilizer used as well as to the time the fertilizer is applied are possible. This provides both environmental and economic benefits [32]. Furthermore, fertilizer management and application is one of the most crucial aspects affecting overall GHG emissions from the biomass production process [19]. GHG emissions from the use of nitrogen fertilizers for biomass production are influenced by two factors. The first is upstream emissions from the production of synthetic fertilizers. It should be noted that upstream emissions differ significantly depending on the chosen nitrogen fertilizer. Selecting a particular nitrogen fertilizer is therefore a promising method of decreasing emissions from the biomass production process [19]. The second important factor involves nitrous oxide emissions from the application of nitrogen fertilizer in agricultural systems. These emissions, often referred to as direct emissions or field emissions, are influenced by a variety of factors, namely climatic and regional aspects as well as spatial aspects such as the technique used for fertilizer application (e.g. especially for organic fertilizer). Their quantification therefore requires exact knowledge of the specific parameters of the site in question. However, a number of simplified approaches for the estimation of nitrous oxide emissions can be found in literature. The Intergovernmental Panel on Climate Change for example provides the methodology for a simplified calculation approach assuming approximately 1 % of the introduced nitrogen to be converted into nitrous oxide [16] Sustainable production of feedstock for the provision of flexible bioenergy therefore requires optimization of fertilizer management, not only of the economic but also environmental aspects.

Biodiversity

The literature available today does not allow for generalized statements on the impact of energy crop production on biodiversity. This impact is site specific and depends on the general characteristics of the cropping system. It can be stated, however, that the cultivation of energy crops provides both opportunities and risks for biodiversity. Positive effects are possible if the production of energy crops leads to improvements in the management system of the area compared to its initial state. As with the aforementioned environmental impacts, these potential positive effects are primarily influenced by the agricultural management system in place and only secondly by the specific type of crops cultivated. Consequently, a high diversity of crops and intelligent management of the crop rotation system with mixed culture species (including a reduction in the use of fertilizers and pesticides) can possibly contribute to increased biodiversity at the site [32]. This might lead to the conclusion that the use of crop rotation systems should not only be optimized in regards to yield increases, but also to achieve the greatest possible contribution to the enrichment of cultural landscapes in terms of different cultural groups of species and cultivation periods [32]. In addition, the use of wastes, residues or landscaping materials (e.g. in the biogas process) in particular can contribute to the protection of valuable habitats and the conservation of biodiversity. The cultivation of energy crops on agricultural land can comprise significant risks to biodiversity. In particular, the intensification of agricultural production systems and the cultivation of monocultures (e.g. due to the resulting increased use of pesticides) can lead to a serious decline of habitats and a significant loss of species. The increasing global demand for biomass for bioenergy and for industrial purposes results in additional pressure on natural areas. In this context, the cultivation of non-native and invasive species presents an additional risk for the conservation of biodiversity. Agricultural areas are the habitat of a variety of organisms. The cultivation of substrates for flexible bioenergy production provides, under consideration of the various guidelines and the existing legal framework for environmental sustainable production, a number of options and opportunities to increase biodiversity. However, the consideration of good agricultural practice and the cultivation in regionally appropriate and meaningful crop rotation systems is always a prerequisite for the cultivation of substrates.

3.3.2 Sustainability Certification

The previously mentioned environmental aspects of biomass production for the provision of flexible bioenergy have been described under the precondition of an existing and functioning governmental framework which addresses the sensitive aspects of agricultural production systems. Unfortunately, such a framework is not in existence in all parts of the world. Considering the fact that biomass feedstocks are increasingly becoming globally traded commodities this could lead to additional problems as far as the sustainable production of (especially) imported biomass is concerned. In this context, the use of liquid biofuels for transport purposes in particular has been discussed intensively within the recent years. As a result of this ongoing debate, the European Commission introduced the Renewable Energy Directive including a set of mandatory sustainability criteria as part of an EU sustainability scheme for biofuels and bioliquids [5]. Currently, these criteria only apply to a small share of the potential feedstocks for the provision of flexible bioenergy (biomass for liquid biofuel production). However, it seems possible and meaningful to expand these sustainability criteria also to other sectors of biomass and bioenergy production (including flexible bioenergy provision) in the future. For this reason, the current status of the available systems for the sustainability certification of biomass production, including a brief overview of the criteria included in their standards, will be discussed in the following paragraphs.

Most of the available schemes for the certification of a sustainable biomass production follow the set of sustainable criteria included in the Renewable Energy Directive. These criteria can be structured into three main elements. (I) The Directive excludes several land categories, with recognised high biodiversity value, from being used for biomass production. These are: (a) primary forests and other wooded land, (b) areas designated for nature protection or for the protection of rare, threatened or endangered ecosystems or species; (c) highly biodiverse grassland, either natural or non-natural. Biomass should not be produced from material from peatland and land with high carbon stock such as:(a) wetlands, (b) continuously forested areas, (c) land covered by trees higher than 5 m and a canopy cover between 10 % and 30 %. (II) For the biomass feedstock produced in the EU, the cross-compliance rules of the Common Agricultural Policy apply, in accordance with the requirements for good agricultural and environmental conditions. The EU cross compliance regulations refer to preservation of soil and water quality, of biological diversity, careful use of fertilisers and pesticides and air pollution. (III) Third major aspect of the sustainability criteria included in the Renewable Energy Directive is the introduction of mandatory GHG-mitigation thresholds for biofuel technologies compared to a fossil reference value (35 % relative to fossil fuels, to increase to 50 % in 2017 and 60 % in 2018 for new biofuel plants). Furthermore, in 2010 the Commission has published a report to provide EU Member States with recommendations for developing national schemes for solid and gaseous biomass used in electricity, heating and cooling [6].

Based on these criteria a number of certification schemes have been developed over recent years and recognised by the European Commission. Several of these schemes for the agricultural sector address a core set of concerns relating to sustainable farming practices, agrochemical handling and use, safety and health and food traceability, with the sustainability criteria addressing mainly environmental aspects. In addition, a number of new initiatives faced rapid development to establish sustainability certification schemes for biofuels feedstock production in tropical countries, such as palm oil, sugarcane and soybean. The existing certification schemes cover a wide area of objectives from specific sectors (agriculture, forestry, etc.) to specific purposes (fair-trade, environmentally sound cultivation, organic agriculture, etc.). While certification schemes for the agricultural sector (such as IFOAM,[3] GlobalGAP,[4] SAN[5] and FAIR TRADE) have been developed primarily developed to

[3] IFOAM: International Federation of Organic Agriculture Movements.

[4] GlobalGAP: Global Good Agricultural Praxis.

[5] SAN: Sustainable Agriculture Network.

ensure health and safety of given products or develop organic agriculture, forestry standards (such as FSC[6] and PEFC[7]) were set to ensure sustainable management of forests. The following table provides a general overview of different existing certification schemes related to biofuel and bioenergy certification. Depending on their main focus, the detail of the environmental, economic and social sustainability aspects included in the standard of the schemes differs strongly across the different schemes. The Table 3.2 summarises the complexity and the completeness of the environmental criteria included in the main certification schemes for biofuels and bioenergy. It furthermore shows great differences between the schemes with regards the completeness of their standards. Since major aspects for environmental sustainability such as the protection of natural areas are of high relevance for the use of all biomass (not only biomass for bioenergy) the existing certification schemes and initiatives should be developed further with regards to the considered indicators and the markets addressed (food, feed, fibre, fuel etc.).

3.4 Conclusion

The resource basis for bioenergy consists of biomass residues, by-products and waste from different sectors and from energy crop production. The global potential of residues, by-products and waste is in a scale of 5–10 GJ per capita and year, including a wide range of qualities needing additional effort to convert them into bioenergy. Due to the strong influence of a number of parameters (e.g. development of crop yields, population growth, per capita consumption, foreign trade balance etc.) available studies on the global biomass potential of energy crops reach different conclusions.

As a result of the recent discussion about the sustainability of bioenergy, different initiatives and schemes for sustainability certification have been developed and implemented for a number of bioenergy pathways (cf. Table 3.2). The existing schemes differ significantly in regards to the variables they consider and thus in the complexity of their indicators and standards. Since major issues for environmental sustainability, such as the protection of natural areas, are highly relevant to the use of all biomass (not only biomass for bioenergy) in terms of developing a more coherent sustainability framework, the existing certification schemes and initiatives should be developed further by addressing the indicators and markets considered. Furthermore, these certification schemes and initiatives should be made an integral part of international agreements. The future development of these schemes and the possible expansion of their application across the whole agricultural or forestry sector will, additionally, influence the future biomass potentials of energy crops and highlight the actual uncertainties of these future potentials.

[6] FSC: Forest Stewardship Council.

[7] PEFC: Programme for the Endorsement of Forest Certification Schemes.

Table 3.2 Environmental aspects considered by different certification schemes (+ aspect included, – aspect not included) (According to [23])

	EU-RED	GBEP	RSB	ISCC	NTA 8080	RTFO	RSPO	FSC	PEFC	GLOBAL GAP
Environmental impact assessment	–	+	+	+			+	+		+
Good farming practice	+		+		+	+	+	+		+
Site history				+				+		+
Sustainable use of resources	–									
Carbon conservation										
Preservation of above/below ground carbon	+			+	+	+	+			
Land use change	+	+	+	+	+		+	+		
GHG emissions	+	+	+	+	+		+			
Biodiversity conservation										
Biodiversity	+	+	+	+	+	+		+	+	+
Natural habitats, ecosystems	+	+	+		+	+		+	+	+
High conversation value areas	+		+	+		+	+	+		+
Negative, endangered and invasive species	+	+	+			+	+	+	+	
GMO	–		+				+	+		+
Soil conversation										
Soil management, soil protection	+	+	+	+	+	+	+	+	+	+
Residues, wastes, by-products	+		+				+	–		
Use of agrochemicals	+		+	+		+	+		+	
Waste management	+		+	+	+	+	+	+		+
Sustainable water use										
Water rights			+	+						
Water quality	+	+	+		+	+	+			+
Water mangement, conversation			+	+	+	+	+			
Efficient water use		+	+							
Air quality										
Air pollution	+	+	+	+	+	+				
No burning for land clearing/ waste disposal		+			+	+	+			
No burning residues, waste, by-products			+			+				+

In regards to the provision of flexible bioenergy two conclusions can be drawn:

- In regards to the provision of flexible bioenergy, a stronger focus on specific conversion pathways also leads to a higher demand for specific feedstock or additional feedstock preparation. However, this might not change the discussion on sustainable biomass potentials dramatically.
- The biggest challenges relate to biomass potentials from energy crops, which are highly uncertain. With regard to smart bioenergy provision, one possible option could be to increase the flexibility of (existing) larger production units based on energy crops by widening their product portfolio.

Finally, it can be stated that the biggest potential for additional bioenergy provision can be provided by accelerating the transition from traditional to modern bioenergy use. Since almost two-thirds of the current global use of biomass for bioenergy is converted in inefficient processes, such a transformation would also provide a wide range of benefits with regards to social (e.g. health issues as a result of particle emissions), economic (adding value due to a more efficient use of a scarce resource) and environmental (increasing the efficiency of future bioenergy systems might be one important method of reducing the pressure on natural areas) issues.

References

1. A. Bauen, J. Woods, R. Hailes, *Bioelectricity Vision: Achieving 15 % of Electricity from Biomass in OECD Countries by 2020* (WWF International, London, 2004)
2. J.E. Campbell, D.B. Lobell, R.C. Genova, C.B. Field, The global potential of bioenergy on abandoned agriculture lands. Environ. Sci. Technol. **42**(15), 5791–5794 (2008)
3. H.Chum, J.Faaij, G.Moreira, P.Berndes, Bioenergy, in *IPCC Special Report on Renewable Energy Sources and Climate Change Mitigation* (Cambridge University Press, Cambridge, UK/New York, 2011)
4. B. Dessus, B. Devin, F. Pharabod, in *World Potential of Renewable Energies-UNESCO World Solar Summit*, 1993
5. Directive 2009/28/EC of the European Parliament and of the Council of 23 April 2009 on the promotion of the use of energy from renewable sources and amending and subsequently repealing Directives 2001/77/EC and 2003/30/EC. — Official Journal of the European Union
6. European Commission, *Report from the Commission to the Council and the European Parliament on Sustainability Requirements for the Use of Solid and Gaseous Biomass Sources in Electricity, Heating and Cooling SEC(2010)* (European Commission, Brussels, 2010)
7. A. Faaij, in *Global Outlook on Development of Sustainable Biomass Resource Potentials*, Budapest, 2007
8. J. Fargione, J. Hill, D. Tilman, S. Polasky, P. Hawthorne, Land clearing and the biofuel carbon debt. Science **319**(5867), 1235–1238 (2008)
9. G. Fischer, L. Schrattenholzer, Global bioenergy potentials through 2050. Biomass Bioenergy **20**(3), 151–159 (2001)
10. D. Gebel, G. Klingenhagen, *Wieviel Pflanzenschutz brauchen Energiepflanzen?* (Bayer Crop Science Kurier, Monheim, 2008)
11. D.O. Hall, F. Rosillo-Calle, R.H. Williams, J. Woods, *Biomass for Energy: Supply Prospects* (Island Press, Washington, DC, 1993), pp. 593–651

12. M. Hoogwijk, A. Faaij, R. van den Broek, G. Berndes, D. Gielen, W. Turkenburg, Exploration of the ranges of the global potential of biomass for energy. Biomass Bioenergy **25**(2), 119–133 (2003)
13. M.M. Hoogwijk, A.P.C. Faaij, B. Eickhout, B. de Vries, W.C. Turkenburg, Potential of biomass energy out to 2100, for four IPCC SRES land-use scenarios. Biomass Bioenergy **29**(4), 225–257 (2005). Universiteit Utrecht
14. International Energy Agency, *Key World Energy Statistics 2013* (International Energy Agency, Paris, 2013)
15. T. Johansson, K. McCormick, L. Neij, W. Turkenburg, W., The Potentials of Renewable Energy, Thematic Background Paper. International conference for renewable energies, Bonn, 2004
16. K. Paustian, N.H. Ravindranath, A.R. van Amstel, *IPCC Guidelines for National Greenhouse Gas Inventories*. Agriculture, Forestry and Other Land Use, vol. 4 (Intergovernmental Panel on Climate Change, Geneva, 2006)
17. M. Kaltschmitt, H. Hartmann, *Energie aus Biomasse: Grundlagen, Techniken und Verfahren* (Springer, Berlin, 2001). ISBN 9783540648536
18. M. Kaltschmitt, H. Hartmann, H. Hofbauer, *Energie aus Biomasse* (Springer, Heidelberg, 2009)
19. S. Majer, K. Oehmichen, *Approaches for Optimising the Greenhouse Gas Balance of Biodiesel Produced from Rapeseed* (project report) (UFOP, 2010)
20. W. Moomaw, J. Moreira, K. Blok, D. Greene, K. Gregory, T. Jaszay, Technological and economic potential of greenhouse gas emissions reductions, in *Climate Change 2001: Mitigation* (Cambridge University Press, Cambridge, 2001)
21. J. Moreira, Global biomass energy potential. J. Mitig. Adapt. Strat. Glob. Change **11**(2), 313–333 (2006)
22. D. Sauerbeck, Funktionen und Bedeutung der organischen Substanz für die Bodenfruchtbarkeit – ein Überblick, in Bundesministerium für Ernährung, Landwirtschaft und Verbraucherschutz (Hrsg.) *Berichte über Landwirtschaft* (1992), S. 13–29
23. N. Scarlat, J. Dallemand, Recent developments of biofuels/bioenergy sustainability certification: a global overview. Energy Policy **39**, 1630–1646 (2011)
24. F. Scheffer, P. Schachtschabel, H.-P. Blume, *Lehrbuch der Bodenkunde* (Spektrum Akademischer Verlag, Heidelberg/Berlin, 2002)
25. T. Searchinger, R. Heimlich, R.A. Houghton, F. Dong, A. Elobeid, J. Fabiosa, S. Tokgoz, D. Hayes, T. Yu, Use of U.S. croplands for biofuels increases greenhouse gases through emissions from land-use change. Science **319**, 1238–1240 (2008)
26. R.E.H. Sims, R.N. Schock, A. Adegbululgbe, J. Fenhann, I. Konstantinaviciute, W. Moomaw, H.B. Nimir, B. Schlamadinger, J. Torres-Martínez, C. Turner, Y. Uchiyama, S.J.V. Vuori, N. Wamukonya, X. Zhang, Energy supply, in *Climate Change 2007: Mitigation. Contribution of Working Group III to the Fourth Assessment Report of the Intergovernmental Panel on Climate Change*, ed. by B. Metz, O.R. Davidson, P.R. Bosch, R. Dave, L.A. Meyer (Cambridge University Press, Cambridge/New York, 2007)
27. E. Smeets, A. Faaij, Bioenergy potentials from forestry in 2050. Clim. Change 2007 Mitig. **81**, 353–390 (2007)
28. E.M.W. Smeets, A.P.C. Faaij, I.M. Lewandowski, W.C. Turkenburg, A bottom-up assessment and review of global bio-energy potentials to 2050. Prog. Energy Combust. Sci. **33**(1), 56–106 (2007)
29. D. Thrän, K. Bunzel, U. Seyfert, V. Zeller, M. Buchhorn, *DBFZ Report No. 7-Global and Regional Spatial Distribution of Biomass Potentials-Status Quo and Options for Specification* (DBFZ, Leipzig, 2011)
30. D. Thrän, C. Hennig, E. Thiffault, J. Heinimö, A. Onofre, Development of bioenergy trade in four different settings – the role of potential and policies, in *International Bioenergy Trade – History, Status & Outlook on Securing Sustainable Bioenergy Supply, Demand and Markets.* (Springer, Dordrecht/Heidelberg/New York/London, 2013). ISBN 978-94-007-6981-6, S. 65–101

31. D. Thrän, T. Seidenberger, J. Zeddies, R. Offermann, Global biomass potentials – resources, drivers and scenario results. Energy Sustain. Dev. **14**(3), 200–205 (2010)
32. A.Vetter, K. Arnold, *Klima- und Umwelteffekte von Biomethan: Anlagentechnik und Substratauswahl* (Nr. 182: Wuppertal Papers), 2010
33. S. Warneke, M. Overesch, H.-J. Brauckmann, G. Broll, H. Höper, Auswirkungen des Energiepflanzenanbaus und der Düngung mit Gärresten auf den Kohlenstoffgehalt im Boden – erste Modellierungsergebnisse, in *Bodenbiologische Indikatoren für eine nachhaltige Bodennutzung*, Osnabrück, 2008
34. Wissenschaftlicher Beirat der Bundesregierung Globale Umweltveränderungen, *Welt im Wandel – Zukunftsfähige Bioenergie und nachhaltige Landnutzung*, Berlin, 2009
35. J. Wolf, P. Bindraban, J. Luijten, L. Vleeshouwers, Exploratory study on the land area required for global food supply and the potential global production of bioenergy. Agr. Syst. **76**, 841–861 (2003)
36. H. Yamamoto, K. Yamaji, J. Fujino, Evaluation of bioenergy resources with a global land use and energy model formulated with SD technique. Appl. Energy **63**, 101–113 (1999)
37. H. Yamamoto, J. Fujino, K. Yamaji, Evaluation of bioenergy potential with a multi-regional global-land-use-and-energy model. Biomass Bioenergy **21**(3), 185–203 (2001)

Chapter 4
Flexible Power Generation from Solid Biofuels

Andreas Ortwein and Volker Lenz

Abstract Flexible and demand-based production of electricity and heat (combined heat and power – CHP) from solid biomass is an extremely interesting concept for a renewable energy system as the used fuel shows excellent storability. However, conversion and power generation technology limit flexibility for several reasons.

Combined heat and power plants for the production of solid biomass are today designed for base load operation. The most common systems are steam cycles, organic Rankine cycles (ORC) and combinations of gasification and gas engines. Other available technologies include Stirling engines, fuel cells and thermoelectric generators (TEGs). Some technologies are already able to provide flexibility in power production. Extracting turbines, for example, are able to change the power-to-heat ratio of the system. It is possible to increase flexibility by using additional or upgraded units such as heat or gas storages, new steam turbines or new control systems. Potential solutions for increasing flexibility in combined heat and power production from solid biomass are expected to include micro-CHP systems and gasification units with high flexibility and high power-to-heat ratio. Larger plants may show less flexibility due to their thermal inertness (which sometimes has been part of the design, e.g. to stabilize combustion of fuels with low heating values).

4.1 Introduction

Flexible and demand-based production of electricity and heat (combined heat and power – CHP) from solid biomass is an extremely interesting concept for a renewable energy system as the used fuel shows excellent storability, including an existing infrastructure for logistics and pretreatment (e.g. pelletizing) [1]. Nevertheless, this has as of yet not been realized by operating units. The following chapter will therefore analyse the challenges and opportunities presented by this technical development.

A. Ortwein (✉) • V. Lenz
Deutsches Biomasseforschungszentrum gGmbH – DBFZ,
Torgauer Str. 116, 04347 Leipzig, Germany
e-mail: andreas.ortwein@dbfz.de; volker.lenz@dbfz.de

D. Thrän (ed.), *Smart Bioenergy*, DOI 10.1007/978-3-319-16193-8_4

The flexibility of combined heat and power generation from solid biomass relies on three main factors – thermo-chemical conversion process(es), intermediate energy carriers and power generation technologies. For a systematic approach to increasing flexibility, the next section will give an overview on different technologies used for the generation of power through the use of solid biomass, including at least two basic process steps: Within the first stage, one or more thermo-chemical conversion processes take place (Sect. 4.2). The second stage is focused on power generation through thermodynamic cycles or similar processes (Sect. 4.3). Concepts for power generation and the status quo in Germany are discussed in the following chapter (Sects. 4.4 and 4.5).

Using this classification, the existing flexibility and the potential for its improvement will be discussed in the following chapter. The existing flexibility of and possible improvements that can be made to state-of-the-art technologies will also be evaluated (Sect. 4.6). Future concepts will be discussed in Sect. 4.7. Finally, conclusions regarding the frame conditions will be discussed in Sect. 4.8.

4.2 Thermo-chemical Conversion Processes

In general, the process of thermo-chemical conversion of solid biomass includes the following steps [2, 3]:

- pretreatment (e.g. drying)
- pyrolysis
- gasification
- combustion

It should be noted that the product range of different thermo-chemical processes strongly depends not only on the chosen process steps itself, but also on other parameters such as pressure, gas phase and particle residence time as well as the reaction environment (e.g. inert media, hot sand, hydrothermal environment) [2].

Before starting the thermo-chemical conversion process, **pretreatment** of biomass is very common. This may include chipping, grinding, pelletizing, briquetting or washing. Drying is a thermal treatment with temperatures low enough to induce only minor chemical changes but high enough to evaporate moisture contained within the fuel. Higher flexibility for pretreatment and drying processes is important for changing fuels in terms of type and amount. Furthermore, since some of the processes may require electrical or heat power (e.g. pelletizing, drying), these energy consuming units can be included in an energy management system, e.g. for the purpose of grid stabilization.

In **pyrolysis**, biomass is decomposed by thermally activated chemical processes within an inert environment. The main product energy carriers, depending on the process parameters, are either solids with a higher energy density than the original fuel (e.g. by torrefaction or slow pyrolysis), liquids (e.g. pyrolysis oil) or sometimes gases (pyrolysis gases). Since the solid and gaseous products can usually be stored, in terms of output the process does not need to be more flexible. However, in terms of input, the degree of flexibility should be comparable to that of the pretreatment processes. This often requires advanced process technology and process control.

Gasification is the reaction of the fuel (including solid, liquid or gaseous products of previous pyrolysis) to mainly gaseous products with significant heating values (e.g. mixtures of hydrogen and carbon monoxide). There are several gasification processes, including moving bed gasification (sometimes also called fixed bed gasification), fluidized bed gasification and entrained flow gasification. Depending on the specific gasification process, flexibility differs greatly. While moving bed gasifiers require a relatively long time to start-up due to the slow heat-up rate of the reactor lining, fluidized bed gasification involves a more efficient start-up process [4]. However, fluidized bed gasification start-up also requires some time (and energy) to heat up the bed sand. In the process of entrained flow gasification, the start-up time depends on the amount of ash the gasification reactor has been designed for since the design of the reactor hull (e.g. reactor lining vs. cooling jacket) heavily influences the time required to heat-up. Turndown of gasification processes is restricted, in particular for downdraft and fluidized bed gasification. However, operation in part load mode is possible with a range as high as 20–110 % for moving bed and 50–120 % for fluidized bed gasification [4].

In **combustion**, all fuel components are oxidized to the maximum. Commonly used technologies for combustion include different grate firings, fluidized bed combustion and dust firing. For more information on the flexibility of small scale solid biomass combustion technologies, see Chap. 6. Fluidized bed combustion, in particular for larger scale combustion (>1 MW), shows higher load change rates when compared to grate firing. Start-up for fluidized bed combustion usually requires more time and energy than for grate firings. This is mainly due to the required heat-up time of the bed material [5].

In Table 4.1, different thermo-chemical conversion processes are compared in the context of flexible power production. Start-stop-behavior, ramping ability and load range are evaluated for typical examples for the respective technology. Their classification is based on the expected demands for flexibility, e.g. for secondary and tertiary control (see Chap. 2).

4.3 Power Generation Technologies

Power generation technologies transfer thermal or chemical energy into electricity. This may happen via thermodynamic cycles (e.g. Rankine, Stirling or Brayton cycles) or by direct power production (e.g. by thermoelectric or electrochemical effects).

The following classification of these power generation technologies is based on the main energy carrier from the last thermo-chemical conversion process to the power generation unit. They can be based on steam (water or organic, usage of phase conversion enthalpy), chemical energy (fuel gas, synthesis gas, synthetic

Table 4.1 Comparison of thermo-chemical conversion processes for flexible power generation

Process	Main product	Technology Readiness Level[a] (TRL)	Start-stop-behavior	Ramping (load change) ability	Load range (from nominal power)
Combustion	Heat (Flue gas)	9	o	o/+	30–110 %
Gasification	Syngas	9	o/+	+/++	50–110 %
Slow pyrolysis	Charcoal	9	–	o	50–110 %
Torrefaction	Torrefied biomass	7–8	o	o/+	(70–100 %)
Flash pyrolysis	Pyrolysis oil	6–7	o	+	(70–110 %)
Start-stop-behavior:					
–– impossible or very hard					
– many hours					
o few hours					
+ <1 h					
++ minutes					
Ramping ability:					
–– no ramping					
– <10 % per hour					
o 20 % per hour					
+ 1 % per minute					
++ 10 % per minute					

[a]According to EU definitions in Horizon2020, Work Programme 2014–2015, Annex G

fuels) or sensible heat (e.g. in flue gas). In Table 4.2, different intermediate energy carriers for power production from solid biomass and their respective storage technologies are listed. Classification of storage efficiency is based on the comparison to electric power storage (e.g. in batteries), while loading and unloading access is evaluated in consideration of heat and mass transfer as well as available technology.

In Table 4.3, different power generation technologies are compared in terms of Technology Readiness Level (TRL), start-stop-behavior, ramping behavior (see Table 4.1) and electrical efficiency for typical units.

4.3.1 *Technologies Based on Steam Cycles*

In steam cycles, the energy used for phase changes is the main driver for the process. There are different technologies based on steam cycles. They can be different in terms of the medium (usually water or an organic liquid) and the power conversion technology (usually turbine or engine).

In **steam turbines**, water is used as medium within a Rankine cycle. Water is boiled and superheated to temperatures above 500 °C [6]. The superheated steam, typically with a pressure of 20–250 bar, drives an often multi-staged steam turbine [6, 7].

Table 4.2 Intermediate energy carriers from solid biomass for power production technologies

Intermediate energy carrier	Energy density	Storage technology	Technology Readiness Level (TRL) of storage technology	Storage efficiency	Loading and unloading access
Flue gas	<90 kWh/m^3	Heat storage (hot water, phase change material)	9 (for T < 100 °C), 7 (for T > 100 °C)	–/o	Good
Water steam	<100 kWh/m^3	Steam storage	9	–	Very good
Organic steam	Unclear	Steam storage	1–2	(–/o)	Unclear
Synthesis gas	f(p)	Syngas storage	3–5	(+)	Good
Synthetic natural gas (SNG)	f(p)	Natural gas grid	9	+/++	Very good
Liquid synthetic fuels	>1,000 kWh/m^3	Tank storage	9	++	Very good
Storage efficiency:					
-- <30 %					
– <50 %					
o <70 %					
+ <85 %					
++ >85 %					

Table 4.3 Flexibility of different power generation technologies

Power generation technology	Typical electrical power range	Technology Readiness Level (TRL)	Startstop-behavior	Ramping (load change) ability	Electrical efficiency η_{el}
Steam turbine	>1 MW	9	o	++	25–35 %
Organic Rankine Cycle (ORC)	100 kW … 5 MW	9	o	+	15–25 %
Steam engine	<1 MW	9	+	++	10–20 %
Gas turbine	>30 kW	9	++	++	30–40 %
Gas engine	<500 kW	9	++	++	35–45 %
Integrated Gasification Combined Cycle (IGCC)	>10 MW	7	o/+	+/++	40–50 %
Fuel cell	1 kW … 5 MW	7–8	–	+	35–65 %
Stirling	<500 kW	7–8	+	o/+	10–18 %
Externally Fired Gas Turbine (EFGT)	10–500 kW	6–7	+	+	18–25 %
Thermo-electric Generators (TEG)	<1 kW	5–7	++	+	<4 %

In condensing steam turbines, the pressure at the turbine outlet is very low. Usually it is already a vacuum below 0.1 bar or 46 °C, making it impossible to use the heat behind the turbine [7].

In non-condensing or back-pressure turbines, pressure at the turbine outlet is typically above 1 bar or 100 °C, which allows heat utilization [7].

For both types, the turbine can be built as an extraction turbine allowing for changes in heat output and the power-to-heat-ratio. A controllable amount of steam is extracted from the turbine at an intermediate pressure and temperature [7].

Back-pressure turbines are commonly used for plants in the range of 0.5–5 MW_{el}. While extraction turbines are suitable for plants above 5 MW_{el}, condensing type turbines are commonly used for larger plants with significantly more than 25 MW_{el}.

Large coal fired power plants with condensing type turbines have reached overall electrical efficiencies of 46 %, which is close to the theoretical maximum [6]. For biomass fired plants, lower electrical efficiencies of 25–35 % are common (depending on size and turbine technology).

In general, steam turbines have to be considered as a fairly flexible technology for power production. The availability of steam is usually the limiting factor.

Organic Rankine Cycle (ORC) systems are based on the same thermodynamic principle as steam turbines. The working medium is an organic fluid. Heat is usually transferred to the working medium via a thermal oil to prevent cracking of the fluid.

As there is no water vapor in the system, it can run in stand-alone mode without continuous observation by humans. This reduces working costs significantly. However, at the same time, the electrical efficiency is, due to lower temperatures, much lower (maximum of about 25 %; in a general work cycle approx. 15 %).

Although ORC systems show some flexibility due to the possibility of changing the ratio of electrical power to heat output, it is estimated to be not as good as extraction steam turbines. However, they still function relatively well in part load operation [8].

In **steam engines**, just as in steam turbines, water is used the medium. Instead of a turbine, an engine is used for power conversion. The engine shaft is linked to a generator which produces electricity. Steam engines are typical for rather low electrical outputs of few kW to larger outputs ranging in the hundreds. The electrical efficiency of the overall system is approx. 10–15 %. The most common types of steam engines are piston engines and screw engines.

In comparison to steam turbines, steam engines are very flexible in power output, limited mostly by the availability of steam. In part load operation they perform at an acceptable level but can also perform extremely well [9].

4.3.2 Technologies Based on Chemical Conversion

The main driver in chemical conversion based technologies is the reaction enthalpy of the energy carrier, which can be used in thermodynamic cycles or by electrochemical conversion.

Gas turbines are usually based on the Brayton cycle (also: Joule or Joule-Thomson cycle). In the compressor-stage, the air needed is compressed. The gaseous fuel from biogas or gasification of solid fuels is then injected and combusted in a combustion chamber. Temperature and pressure resultingly increase significantly. Expanding gas can drive a turbine conducted by an electrical generator. The efficiency of gas turbines depends on their size. Their efficiency can reach values of up to 35 %. The remaining flue gases have relatively high temperatures. So the remaining hot gases can be used for additional electricity production. Gas turbines can be considered as being very flexible, they are currently in use for flexible power production and can supply full power within minutes or even seconds even to extremely large turbines. The flexibility of gas turbines can be used for solid biomass, in particular in combination with the production of synthetic natural gas (SNG) or biomethane. The use of synthesis gas (high contents of hydrogen and carbon monoxide) can often require storage of the synthesis gas or a gasification process which is as flexible as the gas turbine. For externally fired gas turbines.

The basic principle of **gas engines** is the Otto cycle. For solid biomass, gas engines are often used in conjunction with small or medium scaled gasification systems. The product gas from the gasification process is cleaned and cooled. Gas engines show high flexibility and an acceptable electrical efficiency rate of between 35–45 %, according to the fuel input to the engine [10].

Integrated Gasification Combined Cycle (IGCC) is a system of gasification, gas turbine and steam turbine. It has the potential to achieve high electrical efficiency of up to 50 % (based on the higher heating value) [11]. Due to the combination of gas turbine and steam turbine, there is some flexibility in power-to-heat ratio. Today, cost-efficient IGCCs demand at least 10 MW_{el} due to the complexity of the system.

In general, **fuel cells** make it possible to generate electrical power from chemical power through an electrochemical reaction within a cell [12]. In the context of solid biomass, fuel cells can be used to produce electricity from synthesis gas or hydrogen.

Solid oxide fuel cells (SOFCs), which can use synthesis gas without further shifting, are operated at high temperatures. Thus, their start-up and shut-down behavior is not optimal, because the fuel cell stacks are easily damaged by large temperature differences. Still, they behave well in part load operation and have good flexibility [12]. There are new developments in the field of SOFC, e.g. the use of metallic cathodes, allows for a higher number of thermal cycles.

The rate of efficiency from gas to electricity is in the range of 35–65 % depending on size and system [12].

4.3.3 Technologies Based on Sensible Heat Conversion

In some conversion technologies, only the sensible heat e.g. of flue gas, is used as a driver.

Stirling engines are based on thermodynamic cycles with a gaseous working medium [13], which could be air, helium, hydrogen or others [14]. Stirling engines are classified as Alpha, Beta and Gamma engines depending on the following variables: compression space, expansion space, cooler, heater, regenerator and, if required, displacer piston [15]. Usually, stirling engines provide a relatively constant power output [14]. Thus, they are not suited for a rather flexible power generation as needed for primary control. The use of stirling engines is still possible to provide daily or seasonal flexibility. For gaseous fuels, a theoretical electrical efficiency rate above 40 % is possible, with ar mere realistic rate in the range of 20–25 % [13]. With solid fuels in the small scale up to a few hundred kW, the typical annual efficiency is currently in the range of 10–18 % [10].

Difficult fuels (as e.g. straw) can be used in **externally fired gas turbines (EFGT)** by introducing external combustion with a heat exchanger, which heats the working gas (e.g. air). Due to the additional heat transfer and the material characteristics of the heat exchanger, the electrical efficiency is significantly lower, it has been described to be in the range of 25 % with the potential to reach 30–35 % [10]. Externally fired gas turbines for the use of biomass have been discussed in literature [10, 16].

In **thermo-electric generators (TEGs)**, the so-called Seebeck effect is used to produce electricity [17]. While electrical efficiency is very low (usually <4 %), TEGs have the advantage of having no moving parts and are thus expected to be a robust technology. TEGs have the potential to supply auxiliary devices, such as control systems or measurement equipment, with power as well as to provide black start capability.

4.4 Concepts for Power Generation from Solid Biomass

In general, concepts for power generation from solid biomass are based on one or several thermo-chemical conversion processes combined and one or more power generation technologies. An overview on possible combinations is given in Fig. 4.1. It should be noted that some possible additional intermediates (like thermo-chemically treated solid biofuels or biobased synthetic natural gas – bio-SNG) or processes (like torrefaction) are left out since they are dealt with in separate Chap. 8. In a small number of cases, the conversion process is integrated into the power generation technology, e.g. for turbines for sawdust [10, 18]. Since there are no available processes on the market, this possibility will not be discussed further.

These possible combinations can be classified into state-of-the-art-concepts (e.g. combustion + steam turbine, combustion + ORC turbine, combustion + Stirling engine, gasification + gas engine), technologically available concepts (e.g. combustion + externally fired gas turbine, gasification + gas turbine) and future concepts (e.g. gasification + fuel cell, combustion + thermo-electric generator). An overview of the estimated TRLs for different concepts is given in Fig. 4.2, including typical power ranges for these concepts.

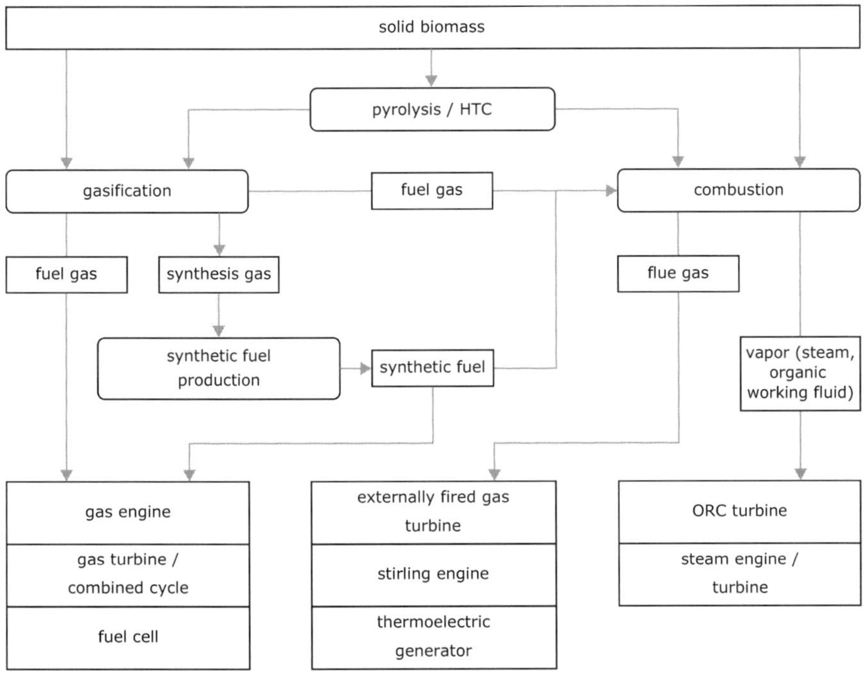

Fig. 4.1 Possible combinations of thermo-chemical conversion processes and power generation technologies, extended from [19] (*HTC* Hydrothermal carbonization, *ORC* Organic Rankine Cycle)

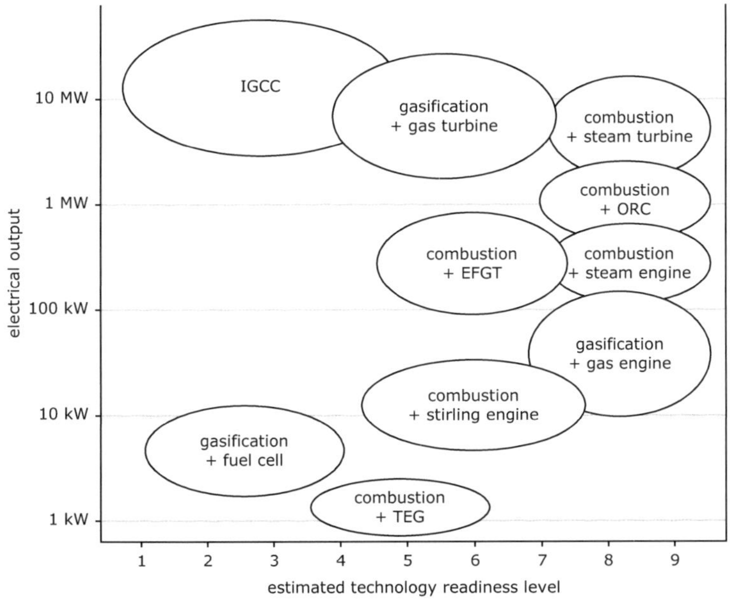

Fig. 4.2 Overview of TRLs for different concepts (*IGCC* Integrated Gasification Combined Cycle, *EFGT* Externally Fired Gas Turbine, *ORC* Organic Rankine Cycle, *TEG* Thermo-electric Generator)

Co-firing concepts, where further technologies could be counted as state-of-the-art, are not considered here since the scope of this work is focusing on renewable energy.

4.5 State of the Art

The most common power plants using solid biomass that are steam based systems, ORC plants and gasification plants with a gas engine (see Fig. 4.3), are to this day usually operated for base load power production due to economical reasons. Some steam based systems are already delivering heat (e.g. process steam) on demand. Furthermore, some of these power plants in Germany are offering tertiary control.[1] Some micro-CHP systems are able to change the power-to-heat ratio [13].

At the moment the number of new installations of steam turbines and ORC turbines has decreased to almost zero. This is caused by increasing biomass prices and decreasing feed-in tariffs. Only the installation of gasifiers together with gas engines has increased rapidly. Due to their low specific electrical nominal power, the total installed electrical power of these gasification units is currently not very high (around 60 MW_{el}).

Some existing plants are already able to provide some flexibility for power production. To achieve further flexibility in existing plants, repowering is necessary.

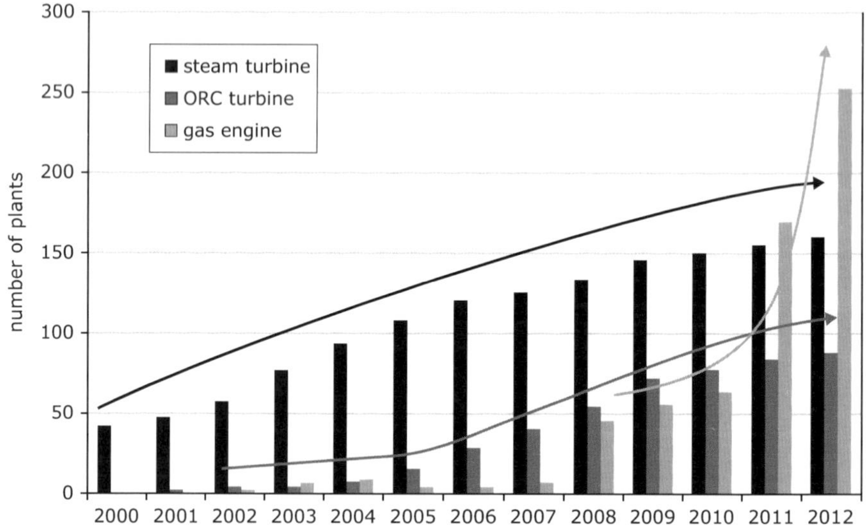

Fig. 4.3 Quantity of steam based systems, ORC plants and gasification plants (According to [24])

[1] Tertiary control is used to stabilize the grid for deviations lasting longer than 15 min.

Furthermore, the feed-in-tariff according to the Renewable Energy Resource Act is currently not offering financial incentives for flexible operation of plants powered by solid biofuels.

4.6 Options for Flexible Power Generation in Existing Plants

Increasing the flexibility of power provision of existing plants powered by solid biofuels can be realized in two main ways. On the one hand, existing equipment within the operation units can be exchanged or improved to increase their flexibility. On the other hand, introducing additional storage options for intermediate energy carriers may provide higher flexibility.

Figure 4.4 shows the basic process schemes for the most relevant technologies for combined heat and power production from solid biomass in Germany. For each

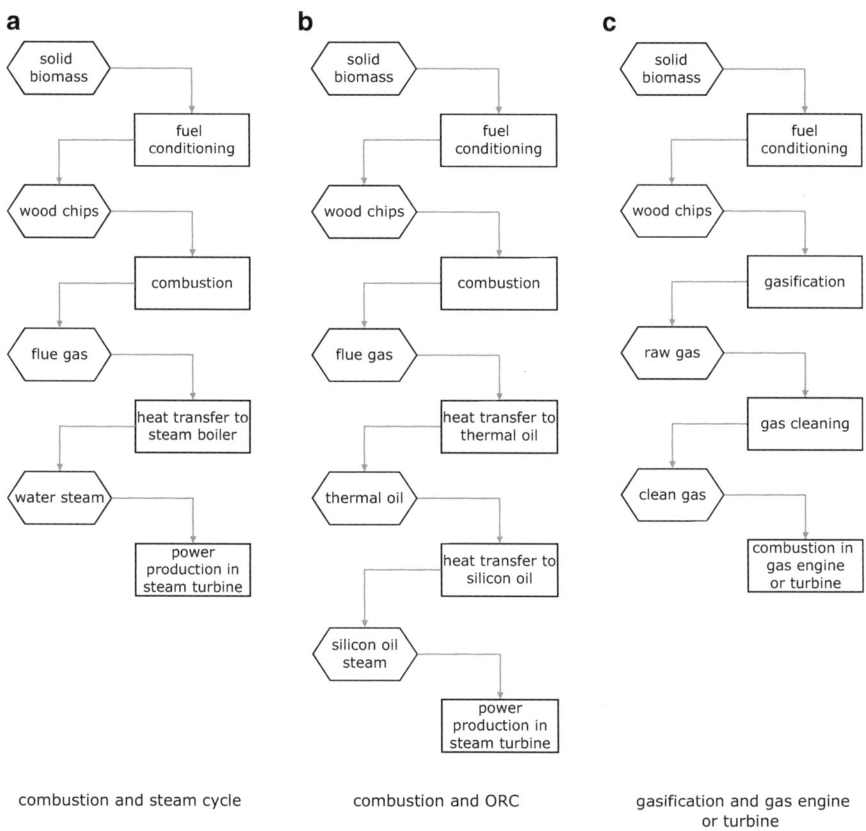

Fig. 4.4 Basic process schemes for the most relevant technologies for combined heat and power production from solid biomass in Germany

technology, the different material streams are listed on the left side due to their importance for storage based solutions. On the right side, the different process units are given that need to be considered for equipment improvement. To increase flexibility the whole chain has to be considered, starting from the last storage option which must offer a high enough capacity to have an effect on flexibility.

4.6.1 Increasing Flexibility of Plant Equipment

Flexibility in existing power plants for solid biomass is strongly dependant on the fuel input and the technology used. For example, some combustion technology is designed for moisture rich fuels (moisture content >40 %). The combustion chamber for such fuels sometimes has very strong brick walls to stabilize combustion. The heat capacity of such walls can limit the flexibility of the combustion process.

Traditionally, ramp rates of steam turbines are below 10 % per minute [20]. Overall flexibility of steam based power plants is much lower. For larger plants, steam bypasses can provide some flexibility within a small modulation range. There are steam cycle based biomass power plants in Germany that are qualified to provide negative tertiary control and that are actively offering control power at the respective markets.

ORC turbines can provide some flexibility by lowering heat transfer from the thermal oil cycle to the silicon oil cycle. The degree of flexibility depends on the heat demands of the overall system.

Gasification systems with gas engines are often already designed to run less than 5,000 h per year due to limited heat demand. If there is already a heat storage, the overall system can provide flexibility.

Steam turbines are offered with a ramp rate of up to 25 % per minute and a minimum turndown rate of 14 % (in certain combined cycle configurations) [21]. If heat provision has to be guaranteed, heat or even steam storage must be installed. Steam storage systems are available but induce losses in efficiency [22]. The overall system, including the combustion unit, can be expected to go down to 30 %. The most common options for increasing flexibility in an existing plant, according to solid biomass power plant operators, are improved control technologies (e.g. software, telecontrol options, enhanced boiler and turbine control).

4.6.2 Storage of Intermediate Energy Carriers

For all existing concepts based on fuel gas, synthesis gas or synthetic fuels as intermediate energy carriers, storage of these intermediates is a general option. Synthetic natural gas can be stored within an existing natural gas grid or in additional natural gas tanks, while even liquefaction is possible. Gases consisting of carbon monoxide and hydrogen and further gaseous components (fuel and synthesis gases) may be stored to improve short-term flexibility, i.e. to provide control power (see Table 4.4).

Table 4.4 Types and time frames of increasing flexible power supply based on solid biofuels

Flexibility time frame	Flexibility type (market)	To balance	Examples for necessary technical adaptations
<15 min	Secondary + tertiary control	Net frequency	Steam bypass or storage (for Steam Cycle Power Plants); gas storage (for gasification + gas engine)
15 min – 12 h	Intraday	Grid schedule optimisation	Advanced control strategies, heat storage
12–24 h	Day ahead	Weather forecast (PV, wind)	Advanced control strategies, heat storage
1–7 days	Day ahead	Macro weather situation (PV, wind)	Long-term heat storage with high capacity
7–90 days	Day ahead	Macro weather situation (wind, PV)	Long-term heat storage with high capacity
90–365 days	Day ahead	Seasonal fluctuations	Increasing efficiency in part load operation e.g. due to constructive changes in combustion chamber

4.7 New Concepts for Flexible Power Generation from Solid Biofuels

4.7.1 General Aspects

For 2050, the European platform on renewable heating and cooling for biomass expects a market potential in the range of 10 GW_{el} for biobased micro-CHP [23]. The technologies already described in Sect. 4.3 are, therefore, expected to be improved and to enter the market within the next 5–20 years. Flexible power production from solid biomass can be achieved through several technologies and varying production plant sizes.

For small scale systems, like micro-CHP, the main concepts for flexible power generation will be

- stabilization of local supply grids,
- minimization of local peaks in power demand or supply ("peak shaving"),
- and stable energy provision for micro grids, e.g. for isolated areas or buildings.

Depending on the concept, control systems for such micro-CHP plants will have to consider grid frequency and voltage as well as smart home aspects or even security of supply. Optimization strategies will focus on economic aspects or on grid stability, depending on the concept. Plant design will strongly depend on local heat demand, including the integration of heat storage and its respective control systems. Typically, the electrical output is expected to be in the range of 1–2 kW_{el} with electrical efficiency of 30–40 %, along with a thermal output of the whole system in the range of 2 kW_{th}. Depending on the integration concept, there will be increased communication between local systems and other entities, e.g. grid control systems, virtual power plants, or alike.

For large buildings and small industry, medium scaled units in the range of 15 up to 250 kW_{el} can play a major role in supplying demand based power and heat. These systems will have to be included in the HVAC (heating, ventilation and air conditioning) as well as in the electrical power control for the building with additional info from the grid. Heat storage and control are also an important aspect for these systems. Electrical efficiency should at least be in the range of 50 %.

New designed fuels (see Chap. 8) can possibly help to improve the technical flexibility of the systems.

4.7.2 *Improvement of Technologically Available Concepts*

Technologically available concepts, that are not state of the art, are usually characterized by a small number of installations and only a few providers. Usually, this leads to a lack of available data concerning economic, environmental and sometimes even technical aspects. From a conceptual approach, the different technologies provide different options of improving flexible power provision:

Combustion and EFGT

Externally fired gas turbines in combination with combustion is a promising option for flexible power generation especially in terms of fuel flexibility. The limiting factor in power production flexibility is the combustion process itself, since there is no steam cycle or the like. Usually, hot exhaust air from the turbine is used as combustion air. By bringing higher flexibility to the combustion process, system and combustion control becomes more challenging.

Gasification and Gas Turbine

Today, most gasification processes are installed in combination with gas engines. The combination of gasification with gas turbines has the potential to achieve a higher rate of electrical efficiency and high flexibility. Depending on the requirements, additional flexibility can be gained by changing the load of the gasification process or by storing syngas as an intermediate energy carrier.

4.7.3 *New Concepts*

In general, gasification is most promising for new concepts for flexible energy production from solid biomass. The product of gasification is a syngas which will give the highest possible amount of flexibility in electrical power production, if properly treated. Since such syngas or (by methanisation) synthetic natural gas can be mixed with biomethane and natural gas, additional degrees of freedom can be reached.

Gasification and Fuel Cells

Although there is already some research on the combination of gasification and fuel cells, the concept has not yet been introduced to the market due to high production costs and strict requirements in gas cleaning. A combined system of gasification and fuel cells, if including syngas and heat storage and an advanced control system, could provide flexibility in several time scales. Further developments indicate, that the reverse usage of fuel cells as electrolysis units is possible. Such a combined biomass-to-gas-to-power-to-gas system could provide control power in the range of −100 % to +100 %.

Hybrid IGCC

In regions with larger potentials of biomass for energy, IGCC can be an option for higher flexibility at high efficiency and nominal power. Since gasification and gas cleaning is the most limiting factor for the flexibility of such an IGCC, hybrid systems using biomethane from the grid and synthesis gas from gasification are promising. If lower electrical power is required within short time frames, reducing biomethane combustion will give a quick response in the gas turbine part, and vice versa.

Synthetic Fuel Production

Flexible production of synthetic fuels via the gasification path has the potential to provide electrical energy on demand by lowering the fuel production (see Chap. 7 – liquid and gaseous biofuels).

By using this approach, gasification and gas cleaning can run on constant power, which might be preferable for some fuels or gasification processes.

4.8 Conclusions

Flexible and demand-based production of electricity and heat from solid biomass is very interesting in the context of a renewable energy system since the used fuel shows excellent storability, including an existing infrastructure for logistics and pretreatment (e.g. pelletizing). However, conversion and power generation technology still restrain higher flexibility for different reasons. For example, some small scale CHP units based on combustion have high requirements on flue gas purity due to the heat exchanger materials and the like, which may limit load change rates. Larger plants may show less flexibility due to their thermal inertness (which sometimes has been part of the design, e.g. to stabilize combustion of fuels with low heating value).

For flexible power generation from solid biomass, there are several options. While some of them are based on already existing and installed technology, there are advanced concepts that will give even higher flexibility. An overview of such concepts is given in Table 4.5. It should be noted, that all estimates of efficiency, load change rates and potential electrical output ranges are strongly dependant on the actual size of the power generation system. For example, the combination of combustion and a steam engine in general is expected to show lower flexibility compared to an IGCC. Nevertheless, a small scale combustion system with a very small steam cycle might have higher load change rates than a large scale IGCC system.

Altogether, future power generation concepts for flexible energy production from solid biomass are required to have high electrical and overall efficiency, high load change rates and a high output range. Together with heat and fast (e.g. electrochemical) power storage, such systems can provide a wide range of flexibility for several applications from primary power control to seasonal variability.

In Table 4.5, the discussed concepts are compared in terms of current status, expected load change rate (see Table 4.1), electrical output range and electrical efficiency. The respective evaluations are a blend of those for thermo-chemical conversion technology, intermediate carrier and power conversion technology. As can be seen in this table, concepts with very high flexibility and efficiency can be expected to rely on gasification as a thermo-chemical conversion process due to the flexible handling of the gaseous intermediate energy carriers.

Table 4.5 Comparison of different concepts for flexible power generation from solid biomass

Power generation concept	Status	Load change rate	Potential electrical output range	Electrical efficiency
Combustion + steam turbine or steam engine	State-of-the-art	o/+	30–110 % (0–110 % with steam storage)	o
Combustion + ORC	State-of-the-art	o/+	0–100 %	–
Combustion + EFGT	Available technology	+	30–110 %	o
Gasification + gas turbine	Available technology	o/+	50–110 % (0–110 % with syngas storage)	+
Gasification + gas engine	State-of-the-art	+	50–110 % (0–110 % with syngas storage)	+
Hybrid IGCC	New concept	++	50–110 % (0–110 % for the gas turbine part)	++
Gasification + fuel cell	New concept	++	−100 % – +100 %	++

While the basic units of these future systems are already available, their complexity requires some further research work. Larger plants can be expected to be installed until 2025 under helpful conditions.

References

1. N. Szarka et al., A novel role for bioenergy: a flexible demand-oriented power-supply. Energy **61**, 18–26 (2013)
2. A. Pollex, A. Ortwein, M. Kaltschmitt, Thermo-chemical conversion of solid biofuels. Biomass Convers. Biorefinery **2**(1), 21–39 (2012)
3. M. Kaltschmitt, H. Hartmann, H. Hofbauer, *Energie aus Biomasse: Grundlagen, Techniken und Verfahren*, 2nd edn. (Springer, Berlin/Heidelberg, 2009)
4. R. Warnecke, Gasification of biomass: comparison of fixed bed and fluidized bed gasifier. Biomass Bioenergy **18**(6), 489–497 (2000)
5. M. Netzer, P. Kolbitsch, Grate firing or bubbling fluidized bed firing – an economic and technological comparison. Presentation at the 4th central European biomass conference, Graz, 2014
6. W. Kalide, H. Sigloch, *Energieumwandlung in Kraft- und Arbeitsmaschinen: Kolbenmaschinen – Strömungsmaschinen – Kraftwerke*, 10th edn. (Hanser, München, 2010)
7. S. van Loo, J. Koppejan (eds.), *The Handbook of Biomass Combustion and Co-firing* (Earthscan, London/Sterling, 2008)
8. M. Van den Broek, B. Vanslambrouck, M. De Paepe, Electricity generation from biomass: Organic rankine cycle versus steam cycle. Presentation at the world bioenergy 2012, Jönköping, Sweden, 2012
9. T. Augustin, Small scale biomass co-generation with modern steam engines. Presented at the IEA-workshop, Copenhagen, Denmark, 2010
10. M. Schmid, *Dezentrale Stromerzeugung mit Feststoffbiomasse; Projektreport* (Ökozentrum Langenbruck, Langenbruck, 2007)
11. M.M.J. Knoope et al., Future technological and economic performance of IGCC and FT production facilities with and without CO_2 capture: combining component based learning curve and bottom-up analysis. Int. J. Greenh. Gas Contr. **16**, 287–310 (2013)
12. P. Kurzweil, *Brennstoffzellentechnik – Grundlagen, Komponenten, Systeme, Anwendungen*, 2nd edn. (Springer, Wiesbaden, 2013)
13. B. Thomas, *Mini-Blockheizkraftwerke: Grundlagen, Gerätetechnik, Betriebsdaten*, 2nd edn. (Vogel, Wurzburg, 2011)
14. B. Kongtragool, S. Wongwises, A review of solar-powered Stirling engines and low temperature differential Stirling engines. Renew. Sustain. Energy Rev. **7**(2), 131–154 (2003)
15. H.I. Onovwiona, V.I. Ugursal, Residential cogeneration systems: review of the current technology. Renew. Sustain. Energy Rev. **10**(5), 389–431 (2006)
16. L. Dong, H. Liu, S. Riffat, Development of small-scale and micro-scale biomass-fuelled CHP systems – a literature review. Appl. Therm. Eng. **29**(11–12), 2119–2126 (2009)
17. S. Riffat, X. Ma, Thermoelectrics: a review of present and potential applications. Appl. Ther. Eng. **23**(8), 913–935 (2003)
18. S.H. Powell, J.T. Hamrick, A wood-fired gas turbine plant, in *Proceedings from the Eighth Annual Industrial Energy Technology Conference*, Houston, 1986, pp. 244–249
19. H. Spliethoff, *Power Generation from Solid Fuels*, 1st edn. (Springer, Berlin/Heidelberg, 2010)
20. C.-K. Weng, A. Ray, Robust wide-range control of steam-electric power plants. IEEE Trans. Contr. Syst. Technol. **5**(1), 74–88 (1997)
21. General Electric Company, D-17 Steam Turbine Fact Sheet (Company Brochure), 2012

22. G. Scholz, *Rohrleitungs- und Apparatebau – Planungshandbuch für Industrie- und Fernwärmeversorgung* (Springer, Berlin/Heidelberg, 2012)
23. B. Sanner et al. Strategic Research and Innovation Agenda for Renewable Heating & Cooling; European Technology Platform on Renewable Heating and Cooling, Brussels, 2013
24. M. Scheftelowitz et al., *Stromerzeugung aus Biomasse 03MAP250* (Zwischenbericht Deutsches Biomasseforschungszentrum, Leipzig, 2013)

Chapter 5
Flexible Power Generation from Biogas

Jan Liebetrau, Jaqueline Daniel-Gromke, and Fabian Jacobi

Abstract The number of plants producing biogas and in particular the technology for converting energy crops and agricultural residues has been increasing substantially in Europe over recent years. The conversion process as well as the utilization of the produced biogas has been designed for a constant operation to allow for maximum capacity utilization. However, a certain degree of flexibility is part of the daily routine operation and in general biogas plants are able to vary their output. Flexibility requires in comparison to steady state operation some additional hardware such as increased gas conversion capacity (e.g. CHP units), a well-adjusted control of gas production and gas storage. This chapter discusses the technical requirements for flexible production and utilization of biogas.

5.1 Introduction

The provision of energy from gaseous biofuels is based on a biological process. Countless species of microorganisms are able to convert organic matter into biogas – mainly methane. Due to the strong support for renewable energies, biogas technology has been able to make substantial progress over recent years. As a result, some several thousand plants are now operating in Europe treating a variety of substrates such as waste water, sewage sludge, manure, energy crops or even municipal solid waste. The flexible energy provision from biogas facilities is currently one of the main technical challenges that need to be overcome to ensure a complete integration of biogas plants into the energy supply system in the future.

J. Liebetrau (✉) • J. Daniel-Gromke • F. Jacobi
Deutsches Biomasseforschungszentrum GmbH – DBFZ,
Torgauer Str. 116, 04347 Leipzig, Germany
e-mail: Jan.Liebetrau@dbfz.de; Jaqueline.Daniel-Gromke@dbfz.de; Fabian.Jacobi@dbfz.de

D. Thrän (ed.), *Smart Bioenergy*, DOI 10.1007/978-3-319-16193-8_5

5.2 Technologies for Generating Power from Gaseous Biofuels

The technology applied in most of the plants (primarily within the agricultural sector) is a straightforward system based on the process of the Continuously Stirred Tank Reactor (CSTR). Due to the fact, that the majority of plants are designed on CSTRs, the chapter mainly discusses this type of digestion technology.

The substrate is usually fed into the digesters by means of pumps (for liquid substrates) and feeding systems for solid matter (e.g. energy crops). The insulated digesters are mainly operated at mesophilic temperatures (37–41 °C) and have rubber domes for gas collection. Due to German regulations, the realized retention times are in most cases more than 100 days with the resulting overall organic loading rates being respectively low. However, some high-rate systems are built rather compact and achieve organic loading rates of 7 $kgVS*m^{-3}*d^{-1}$ and higher. The biogas produced is collected and conditioned (dried and H_2S removed) and either used to produce electricity and heat by means of a combined heat and power unit or upgraded to natural gas quality and fed into the gas grid. Figure 5.1 shows the major components to be found on a biogas plant and displays the main options for gas conversion technologies.

The plants have always been constructed and designed to produce a stable and constant energy output. The biological process in particular shows an optimal performance when operated at a steady state (constant input and output with constant

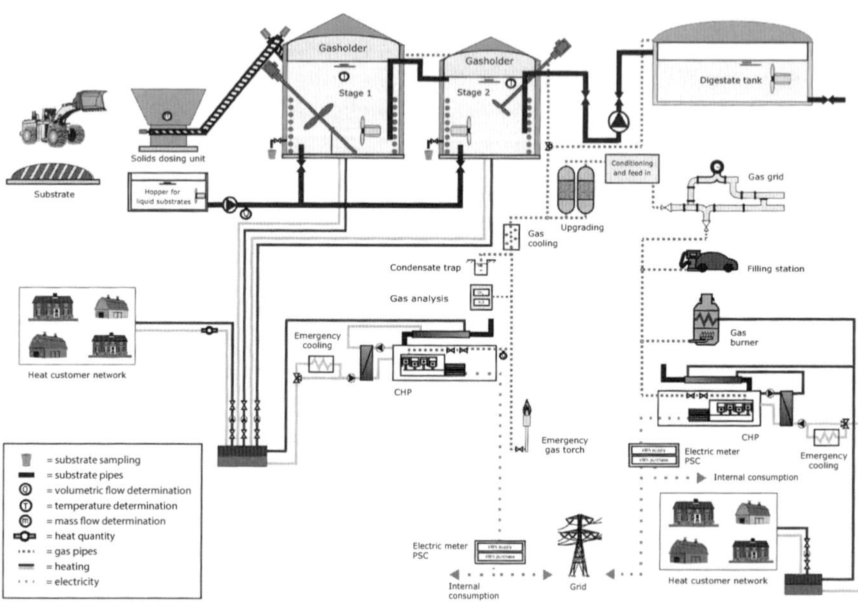

Fig. 5.1 Components of a biogas plant and potential options for gas utilization

process parameters). Due to changing conditions within the energy sector in Germany, biogas plants now have to meet new requirements. These new conditions are mainly related to a flexible electricity supply based on a flexible plant operation.

With the special focus of this book being on the flexibility of the process, the technical components of the overall process will be examined, emphasizing the dynamic behavior of the process. General aspects of the process technology can be found in [12] and [1]. The crucial components or process characteristics defining the flexibility of the process are:

- The type of substrate and substrate supply (characteristics and availability)
- The type of conversion process (plant design/applied technology)
- The gas storage capacity on site
- The type of biogas utilization

Each of the components of the production chain has its own dynamic limitations altogether resulting in the overall flexibility of the process. There are of course other technical components which need to be incorporated into the design if variations in the plant operation are to be considered (e.g. the gas transportation system, the capacity of the feeding technology). In the following, the main process components are described, emphasizing the dynamic characteristics of the process.

5.2.1 Biochemical Conversion Process

Type of Substrate and Substrate Supply

The characteristics and the availability of substrates determine the technology and the plant design for the conversion process. Substrate characteristics such as the content of degradable organics, disturbing matter, trace elements, inhibitory substances and the presence of particulates will have a strong influence on determining the conversion technology. The possible reactor design (e.g. fixed bed, CSTR or dry fermentation) results in a defined dynamic behavior. Within the limitations of the selected reactor design, the degradation rate of a substrate will determine the option to alter the biogas production. The degradation rate of a CSTR can be approximated using first order kinetics. Consequently, the flexibility of the biological part of the process depends on the degradation rate of a given substrate. The Table 5.1 below shows different substrates and a qualitative classification of the degradation rate characteristics.

The Fig. 5.2 visualizes the effect of the degradation rates of substrate in a CSTR on the gas production rate and thus the flexibility of the process itself. The graphs have been obtained from simulations based on experimental data using first order kinetics.

Depending on the degradation rate of the substrates, the process response to the interruption of the process is quite different. Substrates such as sugar beet cause a

Table 5.1 Substrate characteristics

Substrate	DM [%]	VS [%]	Biogas yield $[l_{(STP)}/kgVS]$	Methane content [%]	Degradation rate
Corn silage	33	95	650	52	Medium
Grass silage	35	90	600	53	Medium
Grain-whole-plant silage	33	95	620	53	Medium
Sugar beet silage	23	90	700	52	Fast
Straw	86	90	400	52	Slow
Cattle manure	10	80	380	55	Slow

Source: modified [13]

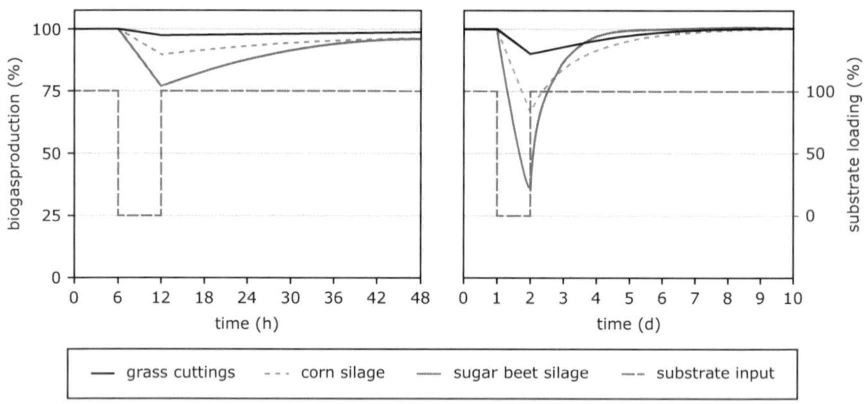

Fig. 5.2 Pattern of biogas production rates for degradation rates found for different substrates and interruption of feeding for 6 and 24 h

rapid decrease in gas production after the interruption of feeding and enable a quick return to full load. By comparison, substrate matter that degrades slowly shows a lower response to a reduction in the input and also responds slowly to continued feeding. The simulation does not imply the potential accumulation of organic acids.

In addition to the reactor design and the kinetics of the substrate degradation, the temporal availability of the substrate will influence the flexibility of the process. Anaerobic digestion processes that treat residues from industrial processes often have limited options to alter the plant throughput in the long term due to a lack of storage capacity and the risk of substrate deterioration during storage. By contrast, plants with digesters treating energy crops use silos for the storage of stabilized substrate and can essentially feed the substrate whenever it is needed. Some plants use a combination of substrates with different storage properties (e.g. liquid manure and silage).

Types of Conversion Process

The conversion of organic matter into biogas can be achieved by means of several fermentation technologies. They include naming only the most important applications for mainly liquid matter: the upflow anaerobic sludge blanket (UASB), the internal circulation (IC) process and the fixed film reactor. For substrates with a higher content of particulates, the continuous stirred tank reactor (CSTR) and the plug flow digester are commonly used, whereas for non-flowing substrates that are impossible to pump, the garage-style percolation systems are available.

Each of the above mentioned conversion processes have their own particular process characteristics which can be described by general kinetic equations of the processes. In addition to process kinetics parameters such as the operational temperature and the substrate characteristics have a strong influence on the kinetics of the overall process. Table 5.2 provides an overview of the main available technologies and their relevant process characteristics. The data have been taken partly from [1].

5.2.2 *Biogas Storage and Utilization*

Options for Gas Storage Within the Process

Gas storage systems are used in the interim between gas production and gas utilization. Generally speaking, gas storage systems for biogas applications are limited to temporarily storing the produced gas over several hours, because the storage of gas in larger quantities is costly. On-site gas can be stored by means of a flexible rubber dome attached to the digesters, by using external gas bag storage or in pressurized tanks. The capacity of the gas storage has so far been designed for short-term flexibility making allowances for fluctuations in gas production or gas utilization. According to a survey on biogas plants, the majority of biogas plants either use rubber domes (app. 47 %) or double shell inflated air domes (37 %) [5].

Gas Conversion/Utilization Technologies

The stationary provision of electricity from biogas can be realized using the following technologies: combined heat and power plants (CHP), gas turbines, fuel cells and gas boilers. In Germany, the CHP utilization is the most widespread technology, since turbines and fuel cells have much higher investment costs and still encounter technological problems when operated with biogas. Gas boilers are rarely used, since possible revenues from the electricity supply are significantly higher than those from the heat supply. Consequently, the sole provision of heat from biogas only prevails in rare cases.

Table 5.2 Anaerobic digestion – process types and flexibility characteristics

Process	Substrates	Total solid content (TS) (within process)	Organic loading rate	Possibility of closing down	Time required for closing down (HRT)	Start up (time from shut down to full load)	Flexible operation: effects on the liability of the process	Partial load behavior
		% wet mass	$kgVS*m^{-3}*d^{-1}$					
UASB (Upflow Anaerobic slugde bed)	Wastewater	Particulates below 1 %	8–15 (COD)	Reduction to 5–10 % of nominal load possible	Equiv. to HRT, (3.3–33 h)	Onefold to threefold of HRT	Risk of process overloading and subsequent failure	Hydraulics influence granula production
Expanded Granular Sludge Bed (EGSB)	Wastewater	Particulates below 1 %	15–25 (COD)	Reduction to 5–10 % of nominal load possible	Equiv. to HRT, (2.2–55) h	Onefold to threefold of HRT	Risk of process overloading and subsequent failure	Hydraulics influence granula production, circulation provides more flexibility than UASB
Fixed film	Wastewater	Particulates below 1 %	10	Reduction to 5–10 % of nominal load possible	Equiv. to HRT, (18–168 h)	Onefold to threefold of HRT	Risk of process overloading	High flexibility and stability under partial load

CSTR	Wastewater, manure, solid matter	6–15 %	Up to 7	Reduction to 5–10 % of nominal load possible	Equiv. to HRT, (app. 15–180 days)	Onefold to threefold of HRT	Risk of process overloading and subsequent failure	Increased substrate utilization, reduction of TS in the process
Plug flow	Organic waste, energy crops	15–25 %	Up to 10	Reduction to 5–10 % of nominal load possible	Equiv. to HRT, (app. 15–25 days)	Onefold to threefold of HRT	Risk of process overloading and subsequent failure	Increased substrate utilization, reduction of TS in the process
Percolation (Garage style)	Structured organic waste	25–40 % Stackable structure for percolation necessary	App. 2.5	Inoculum and/or percolate tank needs to stay active	Equiv. to HRT, (20–30 days)	Onefold to threefold of HRT	Number of active garages can be varied, no impact on processing of single garage	Difficult to handle substrate throughput, steady gas production and substrate structure under partial load

An additional option, which provides a higher degree of flexibility, can be achieved by upgrading the biogas to natural gas quality and subsequently injecting it into the natural gas grid. In this case maximum flexibility is created by totally separating the biogas production and the biogas utilization process. The upgraded biogas is transported in the natural gas grid to the point of gas conversion. This concept allows optimum capacity utilization on the biogas plant site and a constant operation of gas production, gas upgrading and injection into the grid. The upgraded biogas (biomethane) is buffered in the gas grid and can be used on demand. The storage capacity of the gas grid results in a powerful option for decoupling gas production and gas utilization. This enables biomethane-based CHP-units to supply electricity according to the demand and completely independent from the gas production process. This is the major difference compared to on-site CHP-units located directly at biogas plants.

For details on upgrading technologies to biomethane see Chap. 8.

5.2.3 Substrate Availability for Power Generation from Biogas

Plants Based on Energy Crops

A biogas plant treating mainly energy crops includes the substrate storage, the feeding system, digesters, digestate storage, gas conditioning and gas utilization. Substrate storage is necessary, due to the fact that the harvest of energy crops occurs seasonally, although the process needs to be fed continuously. Biogas plants based on energy crops provide numerous flexibility options since the substrate input can be adjusted as required. However, since the biological process, the gas utilization units and the economics of industrial processes all favor a steady operation resulting in high capacity utilization, so far plants have been designed and operated for a constant in- and output.

Manure, Industrial Wastes and Byproducts

Biogas facilities treating mainly manure or waste generally have similar technical components as described above. The main difference is the storage ability and the characteristics of the substrate. Waste or byproducts of industrial processes will occur depending upon the main production process and cannot usually be stabilized and stored over a long period of time. According to this, the flexibility of such plants very much depends on the substrate supply and is usually limited to a short-term storage of the substrate and the storage capacity within the process (e.g. gas storage).

5.3 State of the Art

Around 7,500 biogas plants with an installed capacity of 3,200 MWel were in operation by the end of 2012 (see also Sect. 5.4.) with most of them using energy crops as feedstock. Among the energy crops used, corn silage is the most common with a share of 80 % (mass-related). Grass silage and silages from cereals play a minor role in biogas production. The number of biogas plants solely based on industrial organic waste and separately collected organic waste from municipalities is comparatively low with around 120 plants. A higher number of biogas plants use organic waste as a co-substrate. According to the German Federal Statistical Office approximately 300 plants used organic waste as the main substrate or co-substrate in 2009 [6].

In addition to the biogas plants with on-site electricity generation, nearly 120 biogas plants with upgrading technology to biomethane were in operation by the end of 2012. These plants account for a production capacity of 72,000 m^3 biomethane/hour in 2012 (all given gas volumes refer to dry gas at 0 °C und 101,325 mbar) (Fig. 5.3).

The technology used for biogas production is usually quite simple and robust – the most common type of fermentation process applied within the agricultural sector is the CSTR system (Fig. 5.4).

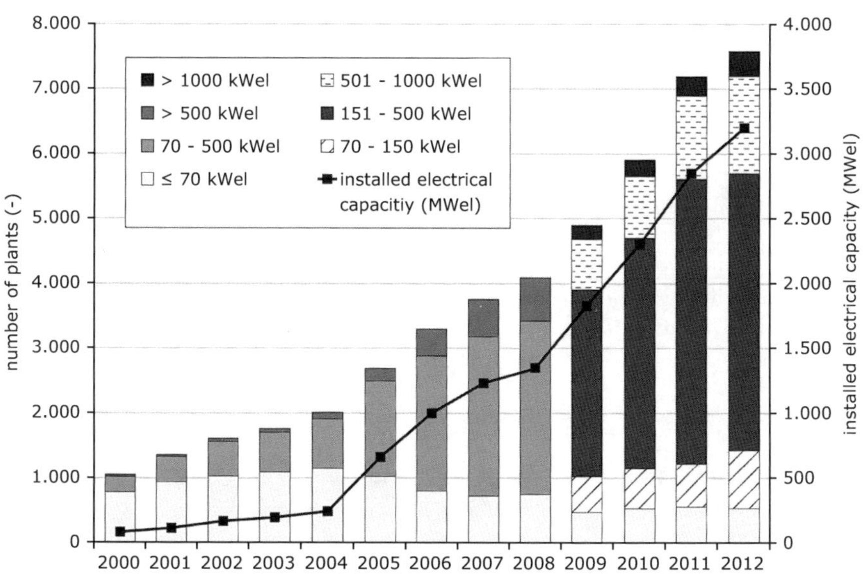

Fig. 5.3 Development of biogas plants in Germany; total number of plants, installed electric capacity ordered by seven plant categories. Biogas upgrading plants, landfill gas plants and biogas plants based on sewage sludge are not included [5]

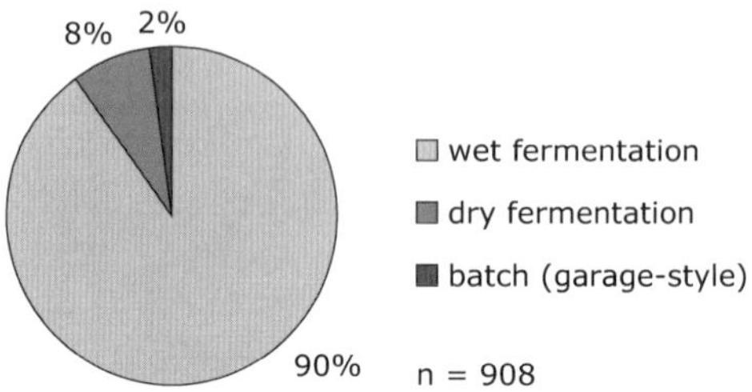

Fig. 5.4 Distribution of applied fermentation processes [5]

Biogas is mainly used for electricity generation (cogeneration of heat and power (CHP)). In addition to on-site electricity generation, micro gas grids are gaining more importance – in order to connect the biogas plant to heat sinks.

- If biogas is used in on-site CHP – on average 1.6 CHP units per plant are installed
- Two-thirds of the CHP units are spark ignition (Gas-Otto) engines (>250 kWel)
- One-third is compression ignition (Diesel engines) (<340 kWel)

The results of a plant operator survey carried out in 2013 [5] showed that the majority (68 %) of plant operators that participated in the questionnaire had not applied for the flexibility bonus from the Renewable Energy Resource Act. From the 21 % of plant operators, which fulfil the preconditions (direct marketing) for the flexibility bonus, only 3 % actually claim it. So far the uncertainties regarding the development of the market and in particular the future of the flexibility tariff have inhibited a fast growing market.

Besides electricity production, heat utilization is also growing in importance. The produced heat can be used in very diverse applications. One of the most common application is the heating of social buildings (70 %) and barns (31 %) from the CHP excess heat. Where drying processes are integrated, over 50 % of them are treating wood, with the drying of digestates showing an increasing share from 5 % in 2010 to 14 % in 2011 [6]. The concept of heat utilization is of substantial importance for the flexibility of the overall process. Since the output of the plant is usually modified in order to meet the demands of the electricity network – the variation of the energy provision still needs to fulfil the requirements of both outlets. Consequently, heat utilization might cause a limitation for the flexibility of the electricity output.

5.4 Options for Flexible Power Generation in Existing Plants

Flexible power generation implies altering the energy output of the plant. Assuming that a plant has a designed capacity that will be met on average, flexibility will mean reduced load times combined with times when the output has to be increased. Given a situation with an average of 10 % down time, the remaining 90 % of the time the plant will have to be operated at an average of 111 % of its design capacity, whereby 60 % of down time would require 250 % of production over the remaining time.

For the same amount of energy to be produced compared to a constant output and a maximum of utilization time, flexible operation certainly requires overcapacity at the gas utilization units as well as times of reduced gas utilization.

Depending on the concept, the additional capacity on the (1) biogas utilisation facilities might require additional capacity in (2) biogas storage systems or even require an alternation of (3) biogas production.

1. **Biogas utilization**

Looking at a biogas plant's flexibility means first of all the installation of overcapacity within the gas utilization facilities. The greater the available overcapacity, the more options are available regarding the difference between the times of maximum output and the times of reduced loads or down time. Further aspects of flexibility are the duration and the amplitude of the power production. The maximum duration of the power supply is defined by both the extent to which the electricity capacity exceeds the equivalent energy provision from gas production and the available gas storage capacity.

2. **Biogas storage system**

The unavoidable rate differences between energy availability from gas production and demand from gas utilization require a gas storage system that balances out the requirements of the two processes. The gas storage capacity has to compensate any deviation of gas production and utilization. Consequently, the gas storage defines the limits of the flexibility, if the gas production process cannot be regulated. Fig. 5.5 shows the available storage capacity of German biogas plants in the agricultural sector according to the survey [5]. The average available gas storage capacity is limited to a maximum of approximately 4 h of gas production under a nominal load (see Fig. 5.5).

In case the gas storage capacity does not suffice to fulfill the requirements of flexibility, the biogas production process itself has to be adjusted. So far, this

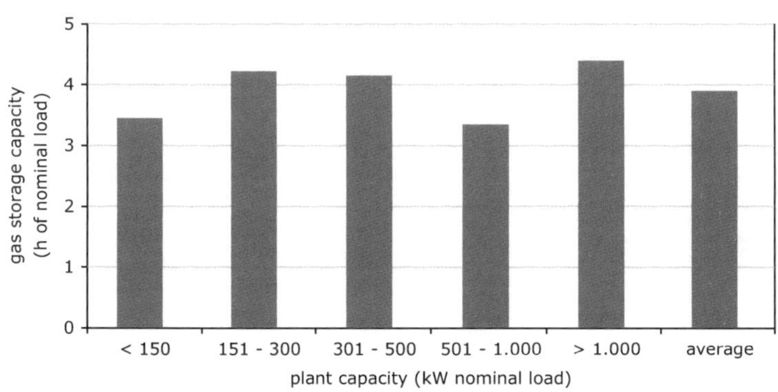

Fig. 5.5 Available gas storage capacity for German biogas plants (considering backup-capacity as well as a correction for temperature, pressure and water vapor saturation)

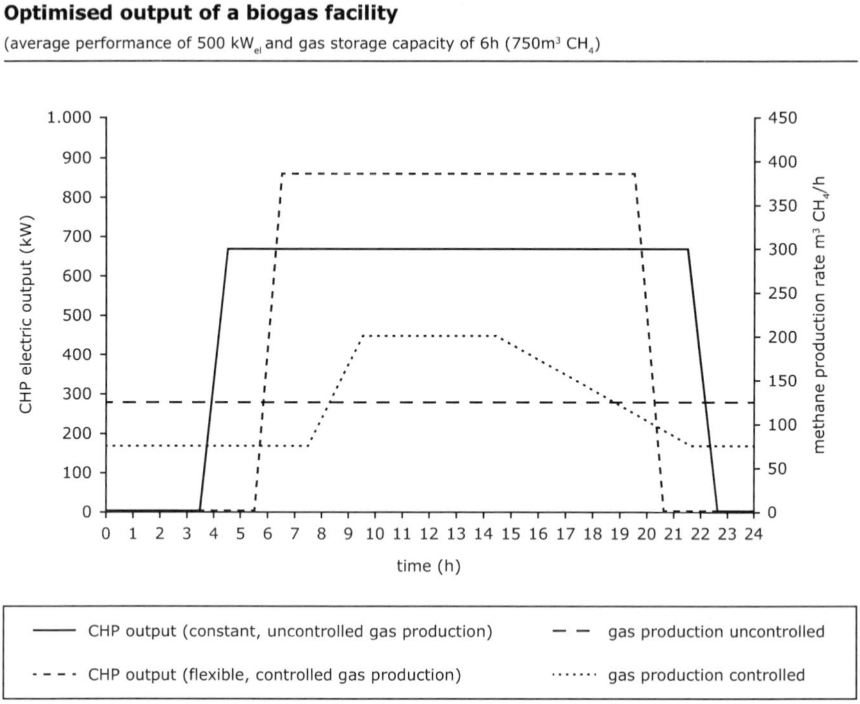

Fig. 5.6 Optimised electricity output from a biogas facility with a controlled gas production rate

has not been done very often and most of the plants do not have the necessary process control technology for the substantial process modifications. However, a regulation of the biological process offers (particularly for long-term modifications) much greater amplitudes rather than relying on gas storage capacities.

Figure 5.6 shows the simulation of a plant, which has an average output of 500 KW (electrical power) and a gas storage for 6 h of gas production (equivalent to the demand of 500 KW el, 750 m^3CH_4). Under scenario one, there is a constant, uncontrolled, i.e. stable gas production, whereas under scenario two there is a flexible controlled gas production, which allows a substantial change of the gas production rate during the day. The CHP operation (assuming that the maximum possible load has to be provided in one go, without interruption) has been adjusted according to the gas production and the filling level of the gas storage. Assuming that both processes need to achieve an average daily output of 500 KW, the regulated process is much more flexible, since it allows more down time (10 h vs. 6 h) and a higher maximum load (857 kW vs. 666 kW).

3. **Biogas production process**

With regard to the biology and the technology of the gas production process there are inherent limitations to flexibility. The anaerobic digestion process requires a stable temperature, which is usually provided by the exhaust heat of the gas utilization units. The activity of the microorganisms requires minimum feeding and the mechanical parts such as pumps and mixers need to be operated frequently in order to avoid malfunctions. Consequently, a complete shutdown of the biogas production for a longer period of time cannot be recommended. Flexible operation with substantial amplitude might require some adjustments to components other than the CHP and the gas storage. Since the throughput of feeding devices, gas transportation pipes or the transformer will also alter according to the given schedule, it might also be necessary to adjust this equipment to the new requirements. Another issue that has to be considered is the simultaneous utilization of the heat and electricity output of the CHP unit. The demand on either side might not fit the demand on the other. To a certain degree compensation can be achieved by using heat storage technology, although this technology will have its limitations regarding the capacity and the duration of storage.

Besides these rather obvious changes in capacity utilization, the speed of the shift and the duration of the down time (or reduced load) are also of importance for the operation of a biogas plant. Depending on the situations described, potential flexibility concepts can be described as follows:

Short term flexibility (reaction time: 5 to 15 mins, duration: up to several hours)

This kind of flexibility can result in a shutdown of the CHP for a short period of time. This would lead to no or minimal changes in the overall plant operation and has little effect on overall capacity utilization. The application is easy to implement. It usually only requires control technology for the CHP units. The limiting factor for the shutdown period is the available gas storage, however this type of flexibility does not require long periods of shutdown. In order to achieve a nominal load, a slight overcapacity is required. It should be mentioned that the CHP needs to be ready for start and stop operation. A temperature control system of the engine might be necessary and the maintenance at the CHP might increase due to stop-and-start operation.

Mid-term flexibility (reaction time: $>$ 15 mins, duration: according to a weekly schedule)

The amplitude and the duration of load alternation are greater. This might be applied in either a shut down for a longer period of time or a continuously reduced load. The alteration is realized within a day or on the following day. In such cases, besides the overcapacity required on the gas utilization side and sufficient gas storage, a feeding management and a correspondent control system for the biological process is necessary. All installations on site have to match the requirements of changes with regard to the throughput. In the case of heat utilization from the CHP exhaust heat, the demand of the heat sink has to be considered and it is possible that additional heat storage technology has to be installed.

Long-term flexibility (reaction time: per season, duration: months)

In this case, the provision of energy is adjusted in the long term. Reasons for such an operation could include seasonal adjustments (e.g. production is ruled by the heat demand of municipal housing), a general increased energy demand in winter, an energy supply for seasonally-operating industries, the utilization of residues from seasonal production processes or long term weather conditions (wind, solar). The limit is again the overcapacity required and the installations on site have to match the changes to operations. Substrate supply has to match the energy output for an adjusted gas production and a control system is necessary in order to avoid critical process conditions during the changing loading rates and in order to be able to adjust the gas production according to the requirements. Looking at the biological process, a shutdown is much easier and safer to accomplish compared to ramping up the process, since the biological process is sensitive to rapid increases in the organic loading rate. Furthermore, increased loading might cause overloading issues and subsequent process failure. Consequently, the plant availability of the biogas production process is more secure through a constant operation, a sudden start up or an increase in the load has a certain risk of process failure. Therefore, the overall process is not recommended for critical increases in loads (e.g. quick startups as an emergency backup for other critical industrial processes).

5.4.1 Critical Components Within the Production Chain

As already described, the major components influencing the flexibility of biogas production and utilization are the substrate characteristics, the biological process, the gas storage and the capacity of the gas utilization. The most sensitive and critical process within that chain is the biological process of gas production. While the CHP units can be shut down and ramped up within minutes, the biological process requires longer time for the same procedure. The rate limitation of the biological system is determined by the type of fermentation technology and the substrate characteristics. The response characteristics of the overall process chain can be improved by an adequate storage capacity of biogas within the process and a harmonized and controlled operation of gas production, gas storage and gas utilization.

5.5 New Concepts for Flexible Power Generation from Biogas

The majority of plants in Germany are operated as continuous stirred tank reactors. These systems are quite simple to operate and easy to deal with. However, due to the large volumes and long retention times, these systems are quite sluggish if it comes to rapid changes. Systems with a high throughput and easily/quickly degradable matter offer a lot more options in terms of flexibility.

New concepts usually focus on the provision of easily degradable matter, which can rapidly be converted into biogas.

The first strategy would be to add easily degradable matter to the substrate portfolio in order to improve the flexibility of a given system. This can be achieved by using substrates such as sugar beet or potent liquid substrates as glycerin from biodiesel production.

Other concepts focus on separating the hydrolysis step and methanisation. Within the hydrolysis step intermediates are formed and separated from the solid material. The methanisation of the acid rich liquid has the advantage of a quick conversion rate of an easily degradable matter and the option of using process types such as the fixed film digester for the second step. Both of them lead to a high degree of flexibility and controllability in biogas production. However, the concept of phase separation is emerging with increased technical effort and has been applied only occasionally [2–4, 8, 11].

A second approach to integrate biogas plants into the energy system is the option of using excess current to increase the methane output. These so-called "power to gas" concepts are based on the assumption that fluctuating renewable energy sources such as wind and photovoltaics produce excess electricity which is available for other processes. One option would be to use this for the production of hydrogen from electrolysis. The hydrogen could be converted to methane by adding carbon dioxide – a major byproduct within the biogas production process [7, 10]. The conversion process can be realized from a thermochemical process in the presence of catalysts or from a biological process. Concepts for the latter include the injection of hydrogen into the process through membranes or the direct injection of the gas [9]; another option could also be a separate fermentation tower for gas conversion.

5.6 Conclusion

So far biogas facilities have been designed for a constant energy output. However, the process can be made far more flexible with only minor changes to plant design. Necessary changes include control systems for gas utilization, adequate gas storage capacity and a sufficient process control system for the biological process. The components technology for a flexible operation is already available, but there is a lack of full scale implementation experience, in particular concerning the knowledge of the interaction of all components and the response of the biological process to load changes.

References

1. W. Bischofsberger, N. Dichtl, K.H. Rosenwinkel, C.F. Seyfried, B. Böhnke, *Anaerobtechnik* (Springer, Berlin, 2005), 718 Seiten. ISBN 978-3-540-06850-1
2. J. Buschmann, G. Busch, Prozessoptimierung in der zweiphasigen/zweistufigen Vergärung fester Biomassen, in *BiogasPOTENZIALE: Erkennen, Erforschen, Erwirtschaften.* Bornimer Agrartechnische Berichte, Bd 79 (Leibniz-Institut für Agrartechnik Potsdam Bornim e.V., Potsdam, 2012), S. 22–33
3. D. Cysneiros, C.J. Banks, S. Heaven, K.-A.G. Karatzas, The role of phase separation and feed cycle length in leach beds coupled to methanogenic reactors for digestion of a solid substrate (Part 2): Hydrolysis, acidification and methanogenesis in a two-phase system. Bioresour. Technol. **102**(16), 7393–7400 (2011)

4. D. Cysneiros, C. Banks, S. Heaven, K.-A.G. Karatzas, The role of phase separation and feed cycle length in leach beds coupled to methanogenic reactors for digestion of a solid substrate (Part 1): Optimisation of reactors' performance. Bioresour. Technol. **103**(1), 56–63 (2012)
5. DBFZ, EEG-Monitoring-Bericht Mai 2013: Stromerzeugung aus Biomasse, Bundesministerium für Umwelt, Naturschutz und Reaktorsicherheit (FKZ 03MAP250) (Leipzig, 2013)
6. DBFZ, Monitoring zur Wirkung des Erneuerbare-Energien-Gesetz (EEG) auf die Entwicklung der Stromerzeugung aus Biomasse. 6. Zwischenbericht, September 2011 (unveröffentlicht). (Leipzig, 2011)
7. G. Gahleitner, Hydrogen from renewable electricity: An international review of power-to-gas pilot plants for stationary applications. Int. J. Hydrog. Energy **38**(5), 2039–2061 (2013)
8. A. Lemmer, S. Zielonka, F. Hahn, J. Lindner, Optimierung der Hydrolyse durch gezielte pH-Wert Steuerung, in *BiogasPOTENZIALE: Erkennen, Erforschen, Erwirtschaften.* Bornimer Agrartechnische Berichte, Bd 79 (Leibniz-Institut für Agrartechnik Potsdam Bornim e.V., Potsdam, 2012), S. 45–57
9. G. Luo, I. Angelidaki, Hollow fiber membrane based H-2 diffusion for efficient in situ biogas upgrading in an anaerobic reactor. Appl. Microbiol. Biotechnol. **97**(8), 3739–3744 (2013). WOS:000317136700045
10. F. Mohseni, M. Magnusson, M. Görling, P. Alvfors, Biogas from renewable electricity – Increasing a climate neutral fuel supply. Energy Solut. Sustain. World **90**(1), 11–16 (2012). Special Issue of International Conference of Applied Energy, ICA2010, Singapore, 21–23 Apr
11. J. Mumme, B. Linke, R. Tölle, Novel upflow anaerobic solid-state (UASS) reactor. Bioresour. Technol. **101**(2), 592–599 (2010)
12. R.E. Speece, *Anaerobic Biotechnology and Odor/Corrosion Control for Municipalities and Industries*, 1st edn. (Archae Press, Nashville, 2008)
13. Kuratorium für Technik und Bauwesen in der Landwirtschaft e.V. (KTBL) Darmstadt (2009) Faustzahlen Biogas; 2. Auflage, S. 94. ISBN 978-3-941583-28-3

Chapter 6
Flexible Heat Provision from Biomass

Volker Lenz and Daniela Thrän

Abstract Heat demand in households always depends on the building, the behavior of the inhabitants, the weather conditions as well many other factors. Therefore, there is always a fluctuating and often not very predictable need for heat. As heating systems have solved this problem for some time now, all heat generators are basically demand-based. Depending on the technology, heat buffering systems are sometimes required. Generally speaking, improved efficiency and low emissions were often achieved in the past by reducing start and stop procedures and applying some kind of base load heat generation. These kind of systems are very commonplace, providing the majority of renewable heat – not only in Germany but also in many other countries. In the future, heat from biomass will have to compare with other renewable heating options and will assume the role of securing heat provision at those times when temperatures fall considerably, when there is limited electricity available in the grid from renewables or when solar thermal systems are not working. This means that the biomass heat generators have to become more flexible in load changes over the total load range without increasing emissions and without significant efficiency losses. Basically, an appropriate design of the conversion system and its conceptual integration will enable a flexible heat supply through solid biomass. The available technologies and concepts for heat supply from solid biomass can be optimized by improved control units, automatic feeding, as well as additional heat storage systems. Consequently, there are a number of options to support the transition to a more renewable-based energy supply, also taking into account better insulation and a fall in the demand for heat in the housing sector. Nevertheless, this transition is more of a vision for decades to come and is still only just emerging in Germany.

V. Lenz (✉)
Deutsches Biomasseforschungszentrum gGmbH – DBFZ,
Torgauer Str. 116, 04347 Leipzig, Germany
e-mail: volker.lenz@dbfz.de

D. Thrän
Department of Bioenergy, Helmholtz Centre for Environmental Research – UFZ,
Permoset Straße 15, 04318 Leipzig, Germany

Deutsches Biomasseforschungszentrum – DBFZ, Torgauer Straße 116,
04347 Leipzig, Germany

Bioenergy Systems, University of Leipzig, Grimmaische Straße 12,
04109 Leipzig, Germany
e-mail: daniela.thraen@ufz.de

D. Thrän (ed.), *Smart Bioenergy*, DOI 10.1007/978-3-319-16193-8_6

6.1 Introduction

Heat is an important final energy use, which can be provided from biomass by different bioenergy carriers and technologies. While the previous chapters included concepts for the combined heat and power provision from biomass, this chapter focuses on the heat-only provision from biomass and the demand for transition in this field. The first section provides an overview of the existing bioenergy carriers and concepts for heat from biomass and future demands for more flexibility are discussed in Sect. 6.2. A more detailed analysis is then provided on the technologies and concepts of heat supply from solid biomass as the most important bioenergy carrier by far for heat provision (Sect. 6.3). Based on the state of the art in Germany (Sect. 6.4) options for a more flexible heat provision in existing plants are then explored (Sect. 6.5) and technical options for new concepts are illustrated (Sect. 6.6).

Finally conclusions are drawn with regard to the more important fields of flexible heat supply during a transition of the energy system and the effects of stronger relationships between power, biofuels and heat markets are demonstrated (Sect. 6.7).

6.2 Heat Supply from Biomass: An Overview and Clarification of "Flexibility"

General Aspects The conversion of biomass to heat is by far the biggest application field of bioenergy worldwide. Even in industrialized countries such as Germany, the heat supply from biomass represents the greatest field of application regarding bioenergy.

Nevertheless, the field of bioenergy will face extensive changes. This assumption is based on recent developments, such as rising costs for oil and natural gas, more stringent regulations in terms of energy saving as well as energy use in building (European Standards, energy saving regulations, laws regarding the minimum amount of energy used from renewable sources) and the increasing amount of partial oversupply of electricity in the grid from renewables. Unlike a combined power and heat supply, which always has to be equipped with a large heat buffer, the heat supply from a flexible operated firing can provide the possibility of optimized overall economic efficiency. In the future, this kind of optimization will become increasingly important due to:

(A) the need for adaptation to a fall in the demand for heat through improved building insulation and
(B) the need for an increased integration of fluctuating, alternative and renewable energy sources.

In most parts of the world a heating system is required for at least some time of the year. However, the allocation of heat usually underlies seasonal fluctuations. On the contrary, domestic hot water (DHW) or process heat is needed throughout the

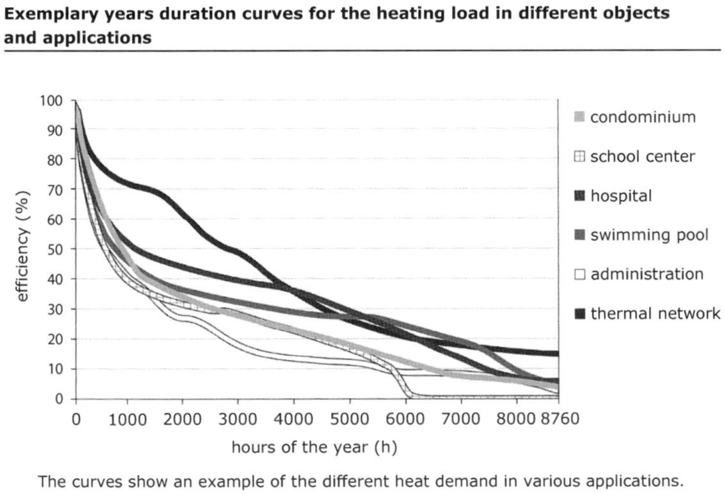

Fig. 6.1 Average heat consumption curves for space heat and DHW in Central Europe [10]

year (see Fig. 6.1). This leads on the one hand to a long-term variation in the average (space) heat demand per day and on the other to a demand for space heat, DHW or process heat that can change rather drastically during the day.

Seasonal fluctuations in demand for heating as well as the possible supply of renewables can only be compensated through heat buffering, which involves tremendous effort and high energy losses or the storage of biomass for demand-oriented conversion. However, short-term fluctuations can partly be balanced through the thermal inertia of heat distribution, user behavior or a well-organized heat buffer size. Even small heat buffer storage tanks can result in investment costs and thermal losses. Therefore, an economic and energetic optimum between the size of the buffer, the frequency of use as well as the flexibility of the boiler has to be found.

Heat Provision Options from Biomass The heat supply from biomass can be based on solid, liquid or gaseous bioenergy sources:

1. Biomethane will be used as a natural gas substitute, obtained through the natural gas distribution system or a gas tank. Furthermore, the same conversion plants are used with the standard start-up and stopping time, extremely small thermal masses and the ability to provide DHW (with a temperature of 70 °C) in less than 1 min. With respect to the great quality of fuel gas used in combined heat and power engines, it is rather unusual in Germany to only use biomethane gas for the purpose of heat production. Globally, biogenic gas is more frequently used for cooking or lighting as opposed to only heat allocation [9].
2. From a current day perspective, the use of biogenic oil as a general source for heat production is not to be expected (excluding niche applications such as bioethanol stoves) [6].

3. Heat supply from solid biofuels can involve manually or automatically-fed furnaces with the most common fuel being wood. Solid fuels have numerous advantages such as almost unlimited storage stability in the right places, ease of storage on site and their associated flammability at all times. Due to these advantages solid biomass seems to be a very promising option for a flexible heat production.

On closer inspection of the technologies available, it is easier to identify their limitations. For example, fireplaces or burn-through wooden log boilers can only stop their heat supply (once ignited), when the entire fuel batch has been fully combusted. Besides which, for a long time the efficiency and the reduction of emissions was the main priority of automatically-fed boilers. Other related goals include long, undisturbed operating phases and only a few starting and stopping cycles. To meet these targets, many combustion chambers were fully faced with fireclay. This however also means reduced ability to adjust the power. With respect to the supply of DHW, the difference between gas and solid biomass furnaces becomes very clear. A gas furnace is able to produce heat immediately according to the user's requirements. However, a solid biomass boiler usually needs a DHW tank to be regarded as an as efficient heat supply. Hence, there is still tremendous scope for research on the flexible and demand-oriented provision of heat from solid biomass, and enormous potential for its optimization. Consequently, this chapter will concentrate on solid as opposed to gaseous or liquid biofuels.

Evaluation Basics for Flexible Heat Provision from Solid Biofuels The flexibility of heat provision technology is defined by the time needed from the first occurrence of a control signal to the stable output of the heating power required. As heat provision from biomass always has to be compared with gas or oil boilers, the time perspectives were oriented to these technologies. Further limitations were presented by the typical inertance found in heating systems, e.g. a central heating system requires a certain amount of time to transfer the hot water from the boiler to the radiators, so a few additional minutes are not really noticed by the customer, whereas a time lag of more than 6 min would be noticeable. With some internal buffers and some forecasting, up to a 30-min time-lag could be acceptable, especially for space heating. Anything longer than this time lag would be considered a problem in most cases with the need for further investment. For a time lag over 6 h, these investments would become considerably high (see Table 6.1).

Table 6.1 An overview regarding the flexibility evaluation depending on the time needed to achieve a stable command variable

Evaluation factor	Symbolic acronym	Time between the control signal and reaching the demanded heating power
Very high	++	Less than 30 s
High	+	More than 30 s and up to 6 min
Medium	0	More than 6 min and up to 30 min
Low	–	More than 30 min and up to 6 h
Very low	––	More than 6 h

6.3 Technologies for Heat Provision from Solid Biomass

In particular wood with different qualities is used within the biogenic solid fuels. Nationally, the amount of wood used to provide heat from biofuels exceeds 99 % [7]. Depending on the regional conditions, other biomass is also used such as waste products from the food industry (e.g. nut shells, seeds, damaged grain, husks, grain strip waste, brewer's spent grain, grape marc and mash from breweries) or different kinds of straw. Especially in developing countries, dung is frequently used for cooking and heating purposes.

With respect to the flexibility of the different plants, the whole conversion chain (fuel to thermal use) has to be considered:

- Type of biomass and biomass quality (water content; dimension; ash content; homogeneity)
- Biomass conversion technology (heat generator)
- Plant concept/operation concept
- Heat storage on site
- Type of heat utilization (among other things internal heat storage of the user)

The single elements are displayed in Fig. 6.2.

All of the elements mentioned have their individual time constant in terms of flow capacity as well as the ability to store intermediates. Furthermore, they also have their own options in terms of an ongoing transition to a flexible heat supply.

6.3.1 Type of Biomass and Biomass Quality

In most cases wood as log wood, chips, pellets or briquettes is used for heating purposes. Agricultural fuels as for example straw or energy grain are used only in very few cases and especially in furnaces with more than 100 kW nominal load. Solid biomass can be converted into heat with the appropriate technology, as long as the gross calorific value of the wet and ash-containing biomass is positive. The water content, the fragmented size and shape of the biomass, but also the ash content and conformity of the fuel will all determine the appropriate technology.

The quality of the fuel affects the storage properties. Table 6.2 shows some of the most important relationships.

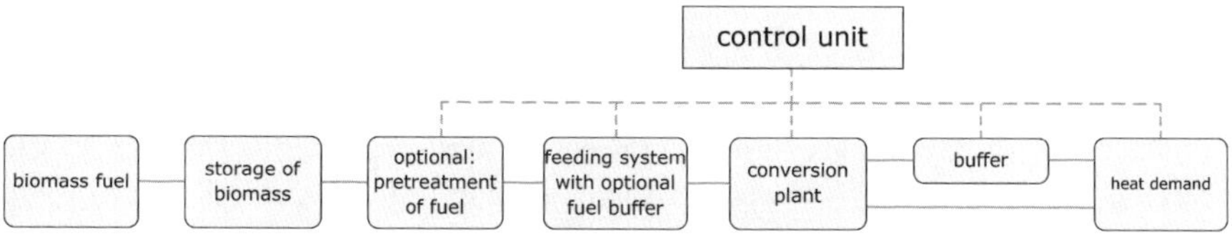

Fig. 6.2 Basic components of heat generation from biomass

Table 6.2 Storage effects of wooden biomass depending on the fuel quality

Parameter	Connection	Effect on storing
Higher water content	A water content higher than 30 % leads to a higher biological activity of the bacteria, causing a reduction of the biomass by heat release. In a huge pile the released heat can hardly be managed, leading to an increased temperature of the pile.	Increased danger of degradation, mass loss or self-ignition
Rougher fragmented size	Without measures to make bulk material more compact, a rougher fragmented size usually causes more openings within the pile. Therefore, the gross caloric value regarding the volume decreases.	More space required to store a certain amount of energy
Higher ash content	The ash bound or attached to the fuel reduces the gross calorific value in terms of the mass and in the end also the gross calorific value in terms of the volume.	More space required to store a certain amount of energy
Higher heterogeneity of the fuel	Heterogeneous fuel dimensions lead to an easy creation of nests or bridging.	Increased danger of transportation disturbances in the fuel storage and in the feeding systems

Table 6.3 Relationships between fuel quality and the flexibility of the operating state of the combustion plant

Parameter	Connection	Effect on flexibility
Higher water content	Additional drying areas within the combustion are needed; more internal buffering to maintain the temperature is necessary	Higher minimum power during operation and slower power adaptation
Rougher fragmented size	A higher amount of fuel is necessary in the combustion plant; a greater variation of the feeding dosage; a greater variation during combustion	Longer reaction time to change the heating power output
Higher ash content	Gross calorific value decreases; afterglow of a comparatively large amount of material in the firebed; partly-burned fuel needs longer to reach the necessary temperature	Difficulties stopping and reigniting the combustion
Higher heterogeneity of the fuel	Fuel-related variation of the primary reaction has to be balanced by internal buffering and sufficient fuel	Longer reaction time to change the heating power output

These factors in addition to the conversion technology used are crucial for safe combustion, low emissions and the time needed to achieve a stable transition of the operating state in terms of heat output. Some of the most important relationships are listed in Table 6.3.

Besides the above-mentioned points (Table 6.3) the fuel and its quality affect the combustion technology. For example, moist fuel is usually burnt in a grate firing to enable an internal drying zone to be integrated. Because of the slower reaction time compared to underfeed furnaces, the reaction times to changes in demand increase significantly without any additional measures.

6.3.2 Biomass Conversion Technology (Heat Generator)

Biomass-based heat supply in industrial countries is usually subjected to requirements other than technological ones (e.g. aesthetics in the living space). Therefore, a number of different technologies are available with the main principles in common of drying, primary conversion and following combustion (in a technical facility) (e.g. underfeed furnaces, site feed firing, grate firing and dropping firing). Each of the technologies has its own advantages and disadvantages regarding the flexibility of heat production (see [5]).

Due to the wide range of biogenic fuels available and the scope of demand for thermal power output, further variations can be added to the large number of technologies available.

For example, the combustion plant can be used as a sole firing or base load boiler. Depending on the type of application, the fireclay lining of the combustion chamber will vary. Therefore, the thermal inertia of the plant will also vary as will the adjustment speed to different heat demands (see Fig. 6.3).

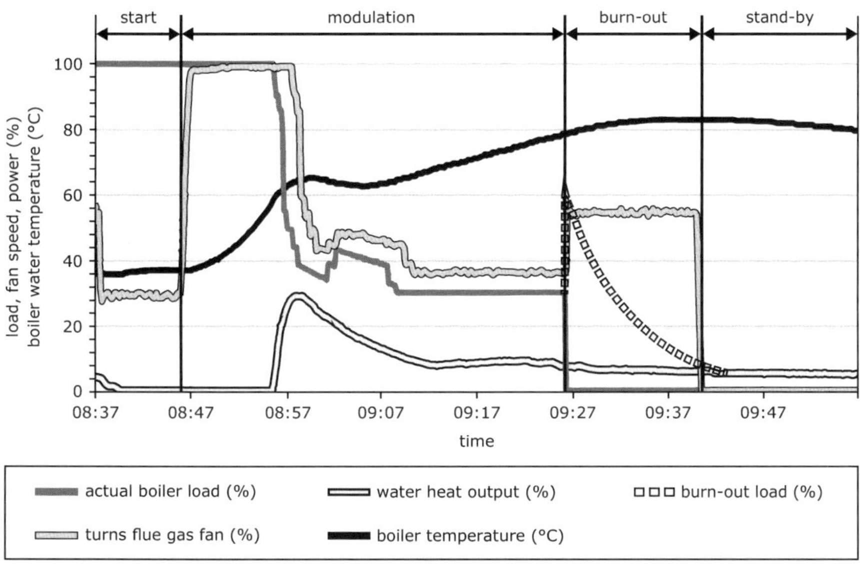

Fig. 6.3 Example of the reaction time of a pellet boiler on start up, load change and burn-out (turns flue gas fan means measure for the speed of the flue gas fan counted in numbers of full turn around compared to the maximum number of turn around possible; burn-out load means thermal power output during the burn-out phase without new fuel input and reduced air input until stop of thermo-chemical conversion compared to the maximum thermal power output)

All of the parameters mentioned lead to a high degree of variation in the time constants needed to evaluate flexibility. Table 6.4 shows some of the most common combustion technologies with their average reaction times.[1]

6.3.3 *Plant Concept/Operation Concept*

The plant concept[2] focuses on different aspects and is usually decided upon at the onset of a project. The central part of the concept is the technology of the biomass conversion unit and whether it is used as a stand-alone heat source or in combination with other heat supply units. The plant/operation concept is influenced by a variety of factors like for example available fuels, space, logistic options, existing heat sources, heat demand structure, personnel.

Based on the heat supply only, some of the basic concepts can be summarized as follows:

Single room fireplace for additional heating. Solid biomass is used in a batch operation mode meaning that the manually-fed fuel is burnt in the combustion chamber one load after the other. Heat is transmitted by radiation through a window and by convection through the walls of the furnace to the surrounding room. The technical development status can vary from one furnace to the other. However, all furnaces have one thing in common. If a combustion load is ignited, it cannot be shut down easily. The heat output of single-room fireplaces usually varies within the range of a few kilowatts (kW). Most of the single-room heaters are not connected to the central heating system and are used only as a separate additional heat source. No buffers or control units are integrated.

Monovalent, mono-fuel central heating system. By means of a biomass boiler heat is produced at a central point. After that, it can be distributed by a suitable heat transfer medium within the building. The plant is designed to guarantee the required heat supply (including DHW supply) throughout the year without an additional heat generation unit. The fuel can be fed to the biomass boiler in two different ways. The log wood boiler is fed in batches with a maximum of two loads per day. An automatically fed boiler (e.g. pellet or woodchip boiler) is fed automatically from the fuel storage as necessary. Most systems have a DHW tank as well as a hot water tank for heating purposes. In most cases, the heat distribution is carried out through hot water but in some cases hot air is used. The combination

[1] According to the wide range of different technologies and constructions, all of the given values are only an average of the total, describing as many of the technologies and constructions without becoming too unspecific.

[2] For the evaluation of the flexibility of heat generation from biomass it is important to understand the difference between conversion technology and the plant/operation concept. The conversion technology is only the heater or the boiler. The plant/operation concept is the conversion technology in addition to all of the components to integrate the heat production into the total heat supply system, e.g. a buffer tank, a second boiler for biomass or even for non-biomass fuels, solar-thermal heat supply and the system control unit.

Table 6.4 Comparison of flexibility for state of the art heat generators

Conversion technology	Fuel quality	Typical thermal power	Start-up time[a]	Range of part load	Time to adjust load	Option to close down	Close down time[b]
Chimneys, stoves and heating inserts	Log wood; w < 20 %	4–12 kW	10–30 min from cold	No	No	Only after full batch	Up to 1 h
Underfeed boiler	Pellets, chips	4–300 kW	10–30 min from cold	30–100 %	5–15 min	Fuel stop and air control to stop the pyrolytic decomposition of the fuel	10–30 min
Grate firing with conversion chambers with fireclay and water cooling	(Wet) wood chips, alternative fuels possible	100 kW to some MW	from 60 min up to 12 h	30–100 % (with very difficult fuels 50 to 100 %)	5–10 min per 10 %- point load change	Either fire keeping without heat output or full cool down program	30–90 min for fire keeping; about 12–24 h to cool down

[a]Start-up time is defined as the time from the signal to start a cooled down boiler from zero power output to the time when the boiler is running constant with a full load power output. Restarting times for a stationary warm boiler can be drastically shorter.

[b]Close down time is defined as the time from the signal to stop the combustion until zero heat output of the boiler. Due to the thermal inertia it could be possible that the boiler still produces heat when the combustion has already finished. In some cases this heat has to be cooled down actively to make the boiler safe. As long as this heat output to the system is running, the close down time is counting.

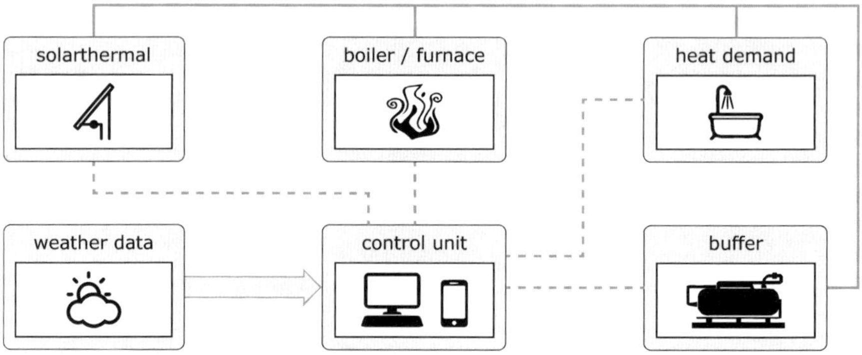

Fig. 6.4 Basic principle of a small-scale biomass mono-fuel heating system combined with a solar-thermal system

of biomass boilers and a solar heating system is quite common. The typical heat output ranges from 4 kW to a few 100 kW. Figure 6.4 illustrates a typical system at present.

Dual-fuel, dual-boiler heating system. To provide enough heat to cover the demand of complex buildings such as office buildings, schools, hospitals or an entire district in the most economical way, different heat generation options can be combined. For example, the system could consist of a biomass boiler to cover the basic load (30–50 % of the possible peak load) and a gas- or oil-fueled boiler to cover the peak load. The comparatively expensive wood boiler is operated using well-priced fuel for most operating hours. The few peaks (on very cold days) can be covered with the far less expensive gas or oil boiler. Hence, the maximum amount of heat provided by the expensive fuel (natural gas or oil) adds up to 30 %. For the overall system, the heat output usually exceeds 300 kW. The heat output of the biomass boiler ranges between 100 kW and some MW. One economic advantage of the concept is when the heat output of the biomass boiler is higher than 300 kW (complete system 600 – some MW).

Besides these three basic concepts, there are also numerous exceptions, new developments and niche applications that are not covered in this book. Nevertheless, the plant/operation concepts mentioned show the main cases that influence flexibility characteristics of heat provision from biomass. The flexibility of these three exemplary concepts is shown in Table 6.5.

6.3.4 Heat Storage on Site

Another decision made within the conceptual design regards the use and size of a heat buffer. Generally speaking, heat can be buffered cost-effectively using different technologies. Depending on the technology used, the costs, storage time and losses can vary. The most important concepts are listed in Table 6.6.

Table 6.5 Comparison of the flexibility of three exemplary state of the art heat provision units for biomass

Heating concept	Surroundings	Start-up time [a]	Range of part load	Time to adjust load	Option to close down	Close down time[b]
Single room fireplace for additional heating	4–12 kW	10–30 min from cold	No	No	Only after full batch	Up to 1 h
Monovalent, mono-fuel central heating system without buffer	4–300 kW	10–30 min from cold	30–100 %	5–15 min	Fuel stop and air control to stop the pyrolytic decomposition of the fuel	10–30 min
Monovalent, mono-fuel central heating system with buffer	4–300 kW	1–5 min	0–100 %	1–5 min	Switch from system heat output to buffering the burn-out heat	Less than 30 s
Dual-fuel, dual-boiler heating system with buffer and system control unit	100 kW to some MW	5–15 min	0–100 %	5–30 min	Either fire keeping without heat output or full cool down program; fossil boiler with fast reaction time	2–5 min

[a]Start up time is defined as the time from the signal to start a cooled down boiler from zero power output to the time when the boiler is running constant with a full load power output. Restarting times for a stationary warm boiler can be drastically shorter.
[b]Close down time is defined as the time from the signal to stop the combustion until zero heat output of the boiler. Due to the thermal inertia it could be possible that the boiler still produces heat when the combustion has already finished. In some cases this heat has to be cooled down actively to make the boiler safe. As long as this heat output to the system is running the close down time is counting.

Table 6.6 Comparison of different heat buffers [3–5]

Technology	Principle	Energy density in kWh/m³	Storage rate in MW	Release rate in MW	Heat loss[a]	Typical size in MWh	Costs in €/kWh
Hot water storage tank	Thermal capacity	<60	<10	<10	10–50 %	<100	0.07–9
Latent heat storage tank	Phase change	<120	<10	<10	10–25 %	<10	9–46
Hot water storage tank with a vacuum insulation	Thermal capacity	<60	<10	<10	2–5 %	<100	28–70

[a]Depending on storage time

Even if heat can be stored more or less easily, Table 6.6 shows that there are significant costs, requirements of space and also energy losses that affect the efficiency of the entire concept. Therefore, the sizing of the buffers is also an important aspect when optimizing a heat supply concept. According to flexibility aspects, the buffer and the buffer integration (heat exchanging capacity) have some additional side conditions in the sizing process. It is most appropriate to use thermal simulation programs that check the efficiency, the costs and flexibility criteria.

6.3.5 *Type of Heat Utilization*

The heat supply requirements are determined by the use of heat and the convenience demand of the user. Usually, DHW should immediately be available at a temperature of 70 °C and a high flow rate is necessary within a short period. However, the user is accustomed to the thermal inertia of the space heating system. Even if heat is emitted from the radiator within a short space of time, it will take up to several hours to heat cold rooms. Therefore, the DHW supply has to be flexible, whereas the response time of the space heating system can add up to 10–30 min. The thermal inertia of the heating system is accepted because the heat demand can be predicted with a high degree of accuracy. The data used to estimate the required amount of heat relies on data such as indoor temperature measurements and outdoor temperature data. Furthermore, a comfortable indoor temperature requires the heat distribution system to show a certain degree of stability (avoiding air draught). Due to slow cooling, an inert mass (internal heat storage) can be beneficial. It leads to a longer uniform temperature distribution of the room in terms of convection and radiation.

The demand for hot water steam is rather common in industrial areas and yet steam heating systems have mostly disappeared by now. The demand for vapor can vary considerably depending on the technical process. Usually, the amount and time of vapor demand are known and the necessary response time is incorporated.

However, the efficiency of the overall system depends on its ability to respond quickly to demand fluctuations. Since steam generation and steam quality require a much more detailed explanation and discussion, they will not be considered at this point.

6.4 State of the Art

Globally, solid biogenic fuel is mostly converted by means of fireplaces for individual rooms. The range of technologies used is exceptionally diverse, covering everything from simple open fireplaces (traditional biomass use) to high-end stoves (modern biomass use) (Fig. 6.5).

Developing and newly industrialized countries still use individual combustion plants to provide heat, which is often their only option. By contrast, industrialized countries, such as Germany, use combustion plants as an additional heat source or even as a luxury good. Pellet stoves are often used as the only heat source for an entire building in developed countries with a low heat demand throughout the year (e.g. Italy). If a higher space heating demand is given (e.g. in Germany or Austria), automatically-fed boilers using wood pellets or chips are used. These boilers are usually able to operate without any assistance and can be compared to an oil or natural gas combustion system. Most of the biomass heat units in the world are however open fireplaces or very cheap stoves or cooking stoves. Even in industrialized countries like Germany however there are still about 14 million single room heaters while there are only about one million boilers installed.

Although there is a lot of discussion about greater efficiency regarding all aspects of the energy supply, there have been no obvious significant changes in the installation numbers of the different heat supply technologies using biomass as a fuel. There has indeed been an increase in the number of pellet boilers installed in Germany every year by a factor of four over the last 10 years. Nevertheless, a total

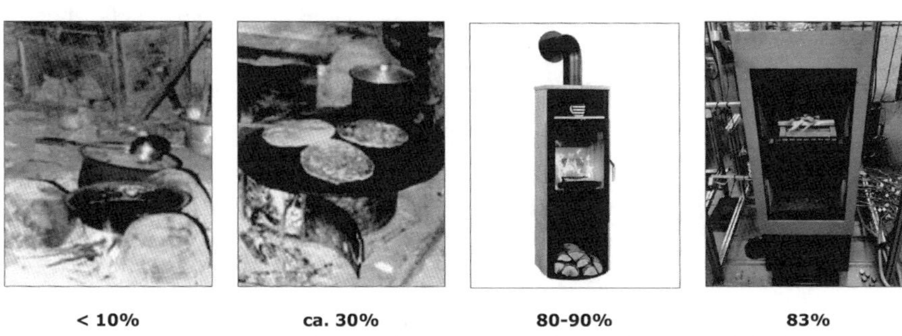

Fig. 6.5 Variety of wood stoves with varying degrees of efficiency (Open fireplace Pakistan; Heating and Cooker Pakistan; Typical stove Germany; High-end stove Germany) [2, 8]

Table 6.7 Evaluation of existing operation flexibilities (for description of symbols see Table 6.1)

	Biomass combustion plant			Plant concept		
System	Start-up	Close down	Thermal power variation	Start-up	Close down	Thermal power variation
Single room fireplace	0	–	Not possible			
Monovalent, mono-fuel heating system without heat buffer	0	+ to 0	+ to 0	0	+ to 0	+
Monovalent, mono-fuel heating system with heat buffer	0	+ to 0	+	+	++	+
Dual-fuel, dual-boiler heating system with heat buffer	– to ––	0 to ––	0	+ to 0	+ to 0	+

of 28,000 units recorded for 2013 is still a very small figure compared to the grand total of 300,000–500,000 new stoves installed per year.

Typical influences on the development of technology are coming from air quality regulations. The combustion of biomass using poor technologies leads to high emissions of e.g. particles. For Europe (including Germany), action has been taking place to reduce the concentration of airborne particles. Therefore, the small-scale combustion regulation [1] in Germany was modified in March 2010 with a significant impact on biomass boilers. Because the boilers will have to achieve very low particle emissions in their everyday operation, there will be improvements to better control the combustion units and greater automation in terms of cleaning, combustion adjustment and emission reductions.

Until now, the flexibility of the combustion units has only been a question of reducing emissions and improving efficiency and competitive advantage by avoiding a heat buffer. In heat supply concepts, the flexibility of the heat supply has in most cases often been a question of buffer size. Hence, little attention has been given to improving the flexibility of the biomass-based heat supply. According to the time values given in Sect. 6.2 and the evaluation criteria of Table 6.1 the state of the art of flexibility of the most common options are listed in Table 6.7.

6.5 Options to Improve Flexible Heat Provision in Existing Plants

Improvements of existing plants are always limited to cost-effectiveness. As the biomass-based heat supply is related to rather high investments, in many cases it can take over ten years to break even. Therefore, adjustments to the installations should not change the basic installations. As a result, in most cases only adjustments to the conversion units and the links between the components are possible, as well as some modifications to the system control.

Single Room Fireplaces as Additional Heating If one considers the flexibility of the heat supply, single room fireplaces fuelled by wood logs (not pellets) are characterized by batch combustion. Therefore, an open combustion can simply be stopped by extinguishing the fire. Otherwise, one would have to wait for the combustion process to complete to ensure safety and low-emissions from the heat supply. Nowadays, the furnaces installed do not allow significant power adjustment. The delay between the ignition of a wood log and the maximum output is also significant. Nevertheless, as an additional heating source, the most critical point in terms of flexibility is a reduction or even stopping of the heat output as even the quickest of central heating systems cannot compensate the overheating of the fireplace in the state of the art fireplace installations. There have been several attempts to equip single-room fireplaces with automatic ventilation flaps. Currently, the main focus has not been on power regulation but on the optimization of the combustion process. Under certain circumstances, the user is able to influence the heat output as well as the combustion period of the firing. E.g. if the allowed amount of wood in the firing is not completely exploited or if the firing is refilled with small amounts of wood. Until now however there are still no sufficient technologies available on the market.

Monovalent, Mono-Fuel Heating Systems As shown in Fig. 6.4, these systems consist of a biomass heat unit that is integrated into a central-heating system. Regarding the heat supply to the heat distributor, it can be distinguished between the response time of the firing unit and that of the overall system, possibly with a heat buffer. Boilers used in combustion plants without a hot water storage tank usually have little internal mass capacity. This leads to a higher reaction rate to power adjustments. Concepts including a hot water storage tank usually feature a higher thermal inertia to stabilize and optimize the combustion process. Therefore, slight differences with respect to firing and strong differences related to the concept of the plant arise and are presented in Table 6.7.

The heat buffer mainly contributes to the shortage of the response time, supplying heat to the distributor. To ensure an optimized operation of both boiler and buffer, the most important facts are: a very specific design of the buffer size, a suitable connection of the boiler and the buffer as well as a working control system. When evaluating a monovalent combustion plant with a heat buffer, three things are fundamental. Firstly, the flexibility of the overall system, secondly, the additional costs for pipe installations and the control system and lastly the additional heat losses of the buffer are essential. These concepts usually provide enough flexibility for common heating and DHW applications. If other technologies, such as solar heat or heat pumps are included in the system, a heat buffer is in any case essential. The heat supply of these systems not only competes against biomass combustion plants, but is also able to provide heat irrespective of the actual demand. In terms of the high initial recognition costs of the solar heating system, the system can only be operated economically if most of the heat produced is used.

To improve the flexibility of the system without changing the boiler, a heat buffer can be integrated if it has not already been installed. Together with an adjusted control unit, the response time to fluctuations in heat demand can be reduced. It has to be recognized that light boilers typically operating without a heat buffer are usually not made to operate at full loads for a long time. Therefore, it is important not to oversize the buffer, to avoid increased abrasion. For a typical house, additional costs including installation are about 1,500–2,500 € for the buffer and the control unit. Besides improved emissions, there could also be an increase in annual efficiency of up to 5 %-points.

To continue to operate a monovalent, mono-fuel biomass heating system without a buffer, it can be upgraded with a power-to-heat unit. With this technology, a continuous-flow water heater can be combined with the biomass combustion plant. The integrated control system will recognize whether or not the heat demand is covered by the electricity or the biomass combustion unit or both. The electrical heater and the control system add further installation costs of about 1,000 € for a typical house. As the electrical heater can start very quickly, the start-up time to deliver heat is drastically reduced. For shut down it is only a help, if a prognosis tool shuts down the biomass boiler in advance and adds the missing heat generation from the electrical heater.

A less technical option to increase the flexibility of an existing system is to change the biomass fuel from a varied fuel (e.g. wood chips) to a more definite fuel (e.g. wood pellets, see Chap. 8). In combination with an adjustment of the boiler and the system control unit, reaction times can then be reduced significantly.

Dual-Fuel, Dual-Boiler Heating Systems The basic idea behind a dual-fuel, dual-boiler heating system consists of the biomass combustion covering most of the full load hours of a year (at least 3,000 h per year). Since a biomass combustion plant is rather expensive, it has to be extremely well-sized, compared to an oil or gas combustion system. The obtainable operational mode also allows the use of biomass of a lower quality, e.g. wood chips with a higher water or bark content. It is mostly grate firings with a fireclay lining of the combustion chamber that are used. These have a high thermal inertia and react very slowly to a change in the demand for heat output. The necessary flexibility of the overall system is usually ensured by the oil or gas combustion plant or a heat buffer. Normally, bi-fuel heating systems with a higher heat output do not operate with a separate hot water storage tank but by using the heat transfer medium within an extensive distribution system (heat pipes between buildings; hydraulic separator). Therefore, the reaction rate is lower compared to a monovalent heating system with a heat buffer (see Table 6.7).

The state of the art of these bi-fuel systems is not very flexible according to the biomass heat generator. If the oil or gas use should be reduced or replaced by other renewable energies such as solar thermal, heat from heat pumps working with renewable energy or even from using the excess of renewable energy from the grid, the flexibility of the system can no longer be guaranteed by the oil or gas component. Therefore, the biomass unit, the system control unit or the buffer possibilities have to improve according to flexible work.

One option for the boiler can be to change the fuel to a more standardized one. A major problem for these systems that economically depend on cheap biomass fuels is the increase in biomass costs. Another option is to add a small and flexible biomass boiler that reacts to all load changes while the basic biomass boiler continues to run with reduced power output at a more or less stable level. If there is enough space for installation, this means additional costs of approx. at least 100,000 € for a 100–150 kW boiler. Additionally, the system control unit has to be improved and in many of the state of the art systems even exchanged. Last but not least, more thermal buffer capacity can be integrated with quick heat exchangers. Again however, costs and additional losses will imply both technical and economical limitations.

General Aspects of Heating Systems The optimization of the system controller is very important. From better forecasting, the contribution from other heat sources such as solar heat or excessive electricity in the grid, a better planning of buffer loading and combustion operation is possible. This results in fewer limitations of the firing flexibility. The predictive presumption of a changing heat demand decreases the response time of the overall system. Therefore, manageable costs for measurement and control devices as well as the connection to the internet arise, allowing a higher flexibility of the heat supply.

Another interesting option is the use of excess heat to dry the fuel during storage (e.g. with high water content), which is partly practiced today. In this way, losses through cooling or fuel residual heat can be used effectively. This leads to a shorter starting phase through dry fuel and a shorter stopping phase.

Table 6.8 shows the flexibilities that could be achieved by the improvements described to existing systems.

Table 6.8 Evaluation of operation flexibility of improved existing heating systems (for description of symbols see Table 6.1)

		Plant concept		
System	Measures	Start-up	Close down	Thermal power variation
Single room fireplace	Automatic air control or fuel amount change by user	0	0	+
Monovalent, mono-fuel heating system without heat buffer	Fuel change to more standardization	0 to +	+ to 0	+
	Integration of a heat buffer	+	++	+
	Integration of an electrical heater with improved control unit	++	+ to 0	+ to ++
Monovalent, mono-fuel heating system with heat buffer	Fuel change to more standardization, better control unit and integration of an electrical heater	++	++	+ to ++
Dual-fuel, dual-boiler heating system with heat buffer	Integration of an additional flexible biomass boiler with standardized fuels; improved control unit with prognosis tool	+	+ to 0	+

6.6 New Concepts for Improved Flexible Heat Generation from Solid Biomass Fuels

In the future, renewable energies alone are supposed to be able to provide almost the entire heat supply. Furthermore, the electricity supply is going to include more energy from renewable sources leading to a temporarily higher fluctuation as well as more favorable prices. Therefore, the priorities for the use of biomass are going to shift, focusing on the heat supply. Energy sources such as solar thermal and geothermal will primarily be used at low and medium power costs and power-to-heat with a current surplus in the power grid. Biomass is going to close the remaining gaps due to a lack of supply or because of economic favorability. Therefore, a future heat provision from biomass will need new concepts.

Single Room Fireplaces Preliminary attempts have been made to further develop single room fireplaces to enable a more flexible operation, even if wood logs are used. Options that enable power adjustment in a single room fireplace using wood logs include: the installation of ventilation flaps, a targeted ventilation duct with a separation of pyrolysis, gasification and combustion air, the integration of burn-off sensors and an automatic control system.

The integration of a water pocket in the firing influences the heat output to the surrounding area, benefiting the heat distribution through the hot water system. Connected to the heat buffer and a centralized heating system, the flexibility of the single room fireplace is similar to that of boilers with a heat buffer. Moreover, a combination with other renewables also becomes possible (e.g. a heat pump with a single room fireplace with a water pocket).

Table 6.9 shows the most important differences in terms of operation flexibility compared to changing investment costs.

Monovalent, Mono-Fuel Biomass Boilers In the future, new installations with this particular heating system concept will only be applied in exceptional cases.

Renewable Heat Station The existing dual-fuel, dual-boiler systems with a base load biomass boiler for cheap biomass fuels together with very flexible oil or gas

Table 6.9 Technical development concepts of single room fireplaces in terms of flexibility and investment costs (for description of symbols see Table 6.1)

Development	Principle	Start	Stop	Alteration of load	Additional costs[a]
Automatic log wood combustion	Wood gasification for single room fireplaces	0	–	+ to 0	100–200 %
Water pocket	Heat output loading the buffer	0	0[b]	+ to 0	50–100 %[c]

[a]Further costs compared to a similar plant without an alteration in %
[b]In terms of a power reduction to 10 % in the room
[c]Without cost for heat buffer or further heat distribution

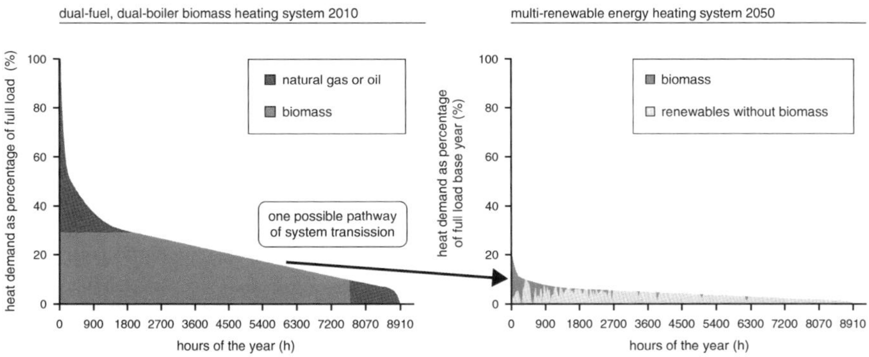

Fig. 6.6 Changing a dual-fuel, dual-boiler system to a renewable heat station

boiler for peak heat demand will become rare in the future. Under the precondition of a strong political target to cover most of the heat supply in the future by renewable energies, there will be a preferred heat use from solar thermal systems, heat pumps and heat from surplus electricity. One option for bioenergy could be to use it only for the missing provision. Due to high fluctuations of the heat sources mentioned, the production of biomass heat must become much more flexible to fill the gap (see Fig. 6.6).

At the nucleus of the renewable heat station is a central control unit that checks all the data of the current and the future heat input of different non-biomass sources as well as the data on demand and the buffer loading. By calculating the missing production, a control signal is sent to the biomass conversion plant. Due to quick changes in the provision from other renewables and differences between real production and prediction, a short response time from the control unit is important to minimize energy costs and losses.

To enable a prioritized use of other renewable energies and to achieve the necessary flexibility, one possible option is to limit the access rights of the biomass combustion for the heat storage volume by the control unit. This does mean however that an optimization of the boiler is necessary in terms of short starting and stopping phases as well as a shorter retention period.

To reach these goals the first step is to develop new ***high-end fuels*** (see Chap. 8). It is hoped that more defined fuels will display better ignition, dosing and burning properties. These properties can result in a smaller firing and less material used to create thermal inertia and therefore shorter response times. A popular method is the transition from wet wood chips to a pre-dried fuel that is easier to use in doses such as wood pellets according to DIN EN 14961–2 and DIN EN 14961–4. Future fuels such as advanced solid biofuels (see Chap. 8) have a higher volumetric energy density, enabling more defined ignition and gasification conditions. Due to a small concentration of volatile compounds or oxygen (from an interrupted primary air supply), it should be easier to terminate the combustion process. To exploit the full potential of these new fuels, an adaptation of the combustion units is necessary.

Compared to premium-wood-pellets, additional costs will arise. These costs will add up to 20–30 % and will be passed on to the customer.

For low thermal outputs in particular, the controllability has to be high. One way of achieving this focuses on the size reduction of the fuel particulates. For example, pellets with a 4 mm diameter and 10 mm length are possible. Furthermore, granulates or pressed fuel balls with a 5 mm diameter are also suitable. This would lead to additional production costs (about 5–10 % of today's production costs), which are feasible.

New boiler developments were necessary not only to exploit the potential of new fuels but also to build units with much greater flexibility according to changes in demand. One trend considers the realization of less inert firing systems with less fireclay or refractory concrete inside the combustion chamber. Besides that, other approaches are also possible.

With the concept of gasification, a primary and secondary separation of the combustion chamber can be carried out more clearly. By means of an optimal integration of a controllable flue gas recirculation with variable water injection, the flue gas production can be altered quickly to a certain extent. In connection with a modular heat exchange concept, allowing an adjustment of the heat exchanger performance, the velocity of the power adjustment will increase. If necessary, an injection of biogas from a network or a cartridge will increase the power for a short period of time, allowing the starting time to be reduced.

To reduce the cold start-up time, the primary reaction zone can be heated by the circulating hot flue gas just like a "small cooling circuit". The primary reaction zone can also be pre-heated by the heat from the hot water storage tank and the primary and combustion air can also be pre-heated.

An improved flexibility of the firing at low combustion capacity is particularly necessary to keep the partial heat buffer costs within an acceptable range.

There are also quite promising options to increase the flexibility of the biomass boilers where the response times required will only be achieved in combination with a buffer. Existing installations have some disadvantages that should be solved within a ***new heat buffering concept.*** The fixed size of the existing heat buffers is a limitation for adjusting the system to the different storage needs over the seasons of a year. In large buildings, cascade connections for buffering systems are already used. This concept can also be applied to smaller plants, but should still involve one closed component. One possible solution could be to integrate vertical separators into the buffer tank, to reduce or to extend the active buffer volume depending on the storage demands.

In larger facilities, a combination of biomass combustion to cover the basic load and a very flexible boiler to cover the peak load can be applied. The combustion chamber of the peak load boiler is lighter and more flexible then the basic load boiler. If necessary, a connected upstream gasification system can be used. In this way it should be possible to avoid seasonal fluctuations. Through adsorption and absorption heat pumps for generating cooling in the summer, the annual demand for heat can be adjusted and organized more efficiently.

Table 6.10 Comparison of the technical components of a newly developed heat station in terms of flexibility and investment costs (for description of symbols see Table 6.1)

Development	Measures	Start-up	Close down	Load changes	Additional costs[a]
High-end fuel	Increase energy density, homogenization	+ to 0	+ to 0	+	10–30 % considering the net fuel costs
New combustion units	Gasification and control system for the amount of gas, pre-heating, modulary heat exchanger	+	+ to 0	+ to ++	30–70 % considering typical boiler price
New heat storage	Internal cascades and adjustment to general requirements	+	+	+	100–200 % considering the usual heat buffer and buffer size
Renewable heat station	Integration of the afore-mentioned measures together with an optimized control unit	+ to ++	+	+ to ++	100–150 % considering the typical boiler price

[a]The additional costs in Table 6.10 are provided compared to actual costs of the non-modified components. From the total additional costs, all of the costs necessary to fulfill more stringent legislation (e.g. emissions) are withdrawn.

Table 6.10 summarizes the parts of a new heating station and the available data. The potential flexibility is also shown.

6.7 Conclusions

In the future, there will be quite a significant number of conventional biomass heat supply systems as base-load heat producers. Depending however on the development of electricity storage systems and on the developments in insulation and efficiency, biomass heating systems will change to peak providers with a much higher flexibility than today. Advanced fuels such as HTC-coal or torrefied wood pellets together with very light and highly adjustable combustion systems will fulfill the needs for heating security while using as many fluctuating renewables as possible such as solar thermal or renewable electricity surplus.

At the same time however an integration of the other renewables mentioned should be reviewed critically. A combination of bioenergy, solar heat and/or ambient heat could be an interesting approach for an integrated heat supply. In such concepts, bioenergy has to provide the remaining load and that will involve overcoming certain technical challenges. One precondition for a more flexible heat supply from biomass is the availability of well-controlled processes with automatic feeding systems and defined solid biofuels. The major developments that are needed are fuel preparation and standardization, pre-gasification and highly adjustable light conversion systems together with modern intelligent system regulators.

Due to a change in the interaction between power to heat demand in housing towards power, the use of micro combined heat and power units (see Sect. 4.7) will probably become more and more of an alternative compared to heat generators only. This could be of great relevance, as they could help to stabilize the electricity grid and to improve the integration of renewables into the energy system.

Additionally, with a more renewable energy supply, the opportunity will arise to generate heat at times when the electricity rate is low from surplus renewable electricity. The feasibility of such concepts will strongly depend on the specific frame conditions, i.e. compare the additional costs of the continuous-flow water heater or heating element within the heat storage tank to the fuel costs of the biomass combustion plant.

With regard to the time frame of the transition, the picture in the heat sector has not been very clear so far:

Changes to buildings and their heating systems are rather slow. At the moment, there are exchange rates of 1–2 % for buildings and 2–3 % for heating systems in Germany [11]. Therefore, establishing a higher flexibility of biomass heat generators and renewable heating systems with biomass will be a process that will last some decades. Especially as there are significant additional costs in most cases that will have currently to be borne by the final user without any clear advantages as the prices of CO_2-certificates are very low, gas and oil prices are at a rather stable level and the market does not provide sufficient refinancing for system integration.

On the other hand, electricity heating inserts in existing hot water buffers could be introduced very soon. The installation and implementation of these systems has already been initiated in combination with private photovoltaic systems with power storage. To use the surplus from summer midday hours, the power is directly transferred to heat. As this power is free of charge, it becomes economically feasible very quickly. As the heating system requires a central heat supply and a buffer to integrate the heater in monovalent biomass heating systems, up to 100 % of the total heat demand can be supplied using this option without any major changes. In the existing dual-fuel and dual-boiler systems, the amount is limited to about 20 % as otherwise the flexibility of the installed boilers is too low.

References

1. Bundesregierung: Erste Verordnung zur Durchführung des Bundes-Immissionsschutzgesetzes (Verordnung über kleine und mittlere Feuerungsanlagen - 1.BImSchV. "Verordnung über kleine und mittlere Feuerungsanlagen vom 26. Januar 2010 (BGBl. I S. 38)"
2. Hark GmbH & Co KG, Product-picture
3. A. Hauer, S. Hiebler, M. Reuß, Wärmespeicher. 5., vollständig überarbeitete Auflage. Bonn, 2013
4. Hummelsberger Schlosserei GmbH, Vakuum-Pufferspeicher. Retrieved from http://www.vakuum-pufferspeicher.de

5. M. Kaltschmitt, H. Hartmann, H. Hofbauer (eds.), *Energie aus Biomasse. Grundlagen, Techniken und Verfahren*, 2nd edn. (Springer, Berlin/Heidelberg, 2009)
6. M. Kutzscher, Biobrennstoffe. Lohnt sich das Umrüsten von Ölheizungen? (2012). Retrieved from http://www.biallo.de/finanzen/Energie/biobrennstoffe-lohnt-das-umruesten-von-oelheizungen.php
7. V. Lenz et al., Erneuerbare Energien. BWK Das Energie-Fachmagazin 4/2014 (2014)
8. S. Nienhuys, Research Report on BACIP Wood Stoves for High Mountain Areas. Gilgit, Pakistan (2000)
9. T. Romberg, Regenerative Energien. Dort Hui, hier Pfui. Die Zeit N°03/2010 (2010)
10. Viessmann Werke GmbH & Co. KG, Beispielhafte Jahresdauerkennlinien für die Heizlast in unterschiedlichen Objekten und Anwendungen. Retrieved from http://www.viessmann.de/de/Industrie-Gewerbe/Systembeispiele/Jahresdauerlinien.html
11. J. Weiß, E. Dunkelberg, Erschließbare Energieeinsparpotenziale im Ein- und Zweifamilienhausbestand. (Institut für ökologische Wirtschaftsforschung, Berlin, 2010)

Chapter 7
Liquid and Gaseous Biofuels for the Transport Sector

Franziska Müller-Langer and Marco Klemm

Abstract In regards to a demand-oriented biofuel supply for the transport sector, this chapter considers the most relevant technologies and concepts for the production and supply of the most important liquid and gaseous biofuels and their current status quo. The limits of and opportunities presented by flexible biofuel production are considered. It has to be noted that flexible or part load operation of biofuel plants is not common. This also applies for most engineering plants in the chemical industry. Today biofuel plants are most commonly constructed as multiproduct plants such as bio refineries. Since the most inflexible step has an effect on the general system flexibility, intermediate storage, raw materials and various products are utilized in order to increase the system flexibility. Flexible management (i) of raw material and other input streams such as auxiliaries (reaction media, catalysts) and (ii) of plant operation in terms of main and by-products including the provision of products with high flexibility in application, is much more common than part load. In the article these opportunities are discussed for existing and new biofuel concepts. Furthermore, general issues of costing and environmental impact are considered.

7.1 Introduction

At present the transport sector accounts for half of global mineral oil consumption and nearly 20 % of world energy use. There will also be increased demand for transport fuels in the future. On a global level approx. 116 EJ a^{-1} are expected until 2050; i.e. an increase of about 25 % compared to 2009 (93 EJ a^{-1}) [13]. The total demand for biofuel is expected to account for 27 % of the total transport fuel demand in 2050 [12]. Biofuels are promoted as one of the best means to account for the predicted increase in future consumption in addition to targeting other priorities such as improved efficiency, traffic reduction and relocation, and electro mobility (Chap. 2). Large quantities may be in demand however due to the complex state of

F. Müller-Langer (✉) • M. Klemm
Deutsches Biomasseforschungszentrum GmbH – DBFZ,
Torgauer Str. 116, 04347 Leipzig, Germany
e-mail: Franziska.Mueller-Langer@dbfz.de; marco.klemm@dbfz.de

D. Thrän (ed.), *Smart Bioenergy*, DOI 10.1007/978-3-319-16193-8_7

affairs in regards to the raw material base for biofuels, the uncertainty surrounding biofuels must be taken into consideration (Chap. 3).

In regards to a demand-oriented biofuel supply for the transport sector, this chapter considers the most relevant technologies and concepts for liquid and gaseous biofuels and their current status quo. Furthermore, the limits of and opportunities presented by flexible biofuel production are briefly discussed.

7.2 Technologies

There are various methods to produce liquid and gaseous fuels from biomass. The purpose of biomass conversion is to provide fuels with clearly defined fuel characteristics that meet given fuel quality standards. Depending on the method of biomass conversion there are three main pathways to consider; all of them are part of specific overall concepts that are characterized by different grades of technological complexity and flexibility [14, 20]:

Physico-chemical Conversion Such processes usually use low temperatures and pressure levels. They include the production and treatment of oil and fat containing biomasses into triglyceride biomass (e.g. vegetable and animal fats and oils) and fatty acids. These raw materials are processed further with alcohols through catalyzed trans-/esterification into biodiesel or fatty acid methyl ester (FAME). It is used in pure form in specially adapted vehicles or is blended with diesel.

Biochemical Conversion These processes involve using microorganisms to convert the biomass (usually sugar and starch fractions) into liquid and gaseous fuels. For instance bioethanol is produced by fermenting sugars from starch and sugar biomass. It is applied in pure form in specially adapted vehicles or blended with gasoline, provided that fuel specifications are met. Another method is using biogas resulting from the anaerobic treatment of biogas substrates, which is then upgraded to biomethane and can then be fed into the natural gas grid and e.g. used in natural gas vehicles. Both of these current developments involve the application of special treatment processes (hydrolysis via thermal processes or enzymes) that succeed in breaking down lignocellulosic biomasses and releasing sugars, which can then be fermented into alcohol or digested.

Thermo-chemical Conversion These processes use high temperatures and pressure levels to turn biomass (usually lignocellulosic fractions) via different methods such as torrefaction, pyrolysis or hydrothermal processes into different products (i.e. depending on process conditions usually into solid, liquid and gaseous fractions) that can be either upgraded or further processed, e.g. via gasification (see also Chap. 8). Gasification based process chains include conversion into a raw gas that is then treated and conditioned into a synthetic gas consisting mainly of carbon monoxide and hydrogen. This gas can be processed further into different types of liquid and gaseous fuels via different fuel synthesis and upgrading technologies. Fuels

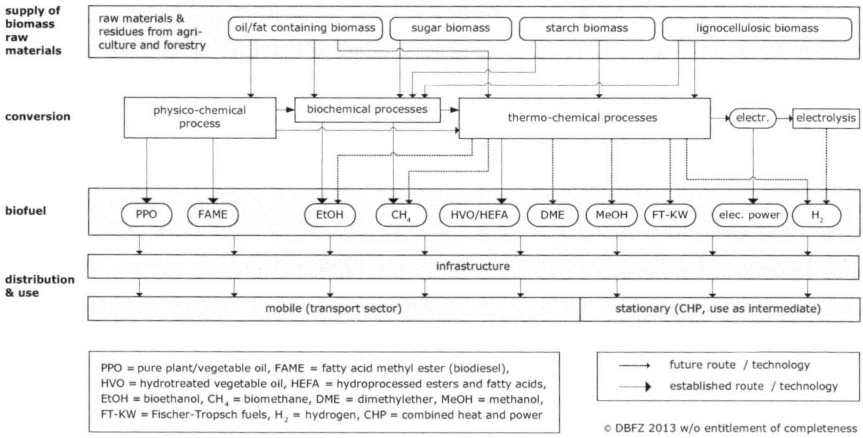

Fig. 7.1 Overview of biofuel options (Adapted from [21])

from this route are then called 'synthetic biofuels'. The most promising liquid synfuel (also BTL, biomass-to-liquids) is e.g. Fischer-Tropsch (FT) fuels due to its favourable fuel properties. Furthermore, alcohols (e.g. ethanol and methanol) can also be produced. Gaseous synfuels are e.g. dimethylether (DME) and biobased synthetic natural gas (Bio-SNG), which is also a form of biomethane and can be similarly used as a natural gas substitute such as biomethane from biogas. Furthermore, available vegetable oils or animal fats from physical-chemical conversion can be treated by hydrotreating processes into so called hydrotreated vegetable oils or esters and fatty acids (HVO/HEFA), a biodiesel with comparably more favourable properties than conventional biodiesel.

A comprehensive overview of the overall supply chains of the most important biofuel options under international discussion is provided in Fig. 7.1.

7.3 Concepts and State of the Art

Usually, within a certain biofuel route (e.g., bioethanol) overall concepts for biofuel production plants are quite different; they cannot be bought off the shelf. In regards to those already in existence, the concepts which have been realised are dependent on the specific local conditions and infrastructure, the equipment provider and certain optimisations through the biofuel production plant operator itself. Each biofuel concept must therefore be considered individually.

Today, biofuel production plants most commonly exist as so called multiproduct plants such as biorefineries. According to [9], material and energy-driven biorefineries can be distinguished. Much of the existing network of biorefineries already has a strong link to biofuel production or energy-driven biorefineries [16]; for instance by-products that are available in addition to the main product biofuel such as fodder, fertiliser, products for further processing in feed, cosmetic and chemical industry. Furthermore, some of the biofuels can also be used in the intermediary stage before further processing in different industry branches (e.g., bioethanol, biomethane, biodydrogen, Fig. 7.1).

According to this, a selection of current and future biofuel options are considered; a summary of their typical technical characteristics, status quo as well as international production rates and capacities is given in Table 7.1.

In addition to the given biofuel capacities in Table 7.1, the development of biofuel production capacities is provided in Fig. 7.2. While biodiesel capacities (mainly based on rape) decreased caused by the development of a policy frame and thus market conditions, bioethanol (based on wheat, rye and sugar beet) slightly increased. In comparison, biomethane (based on different energy crops but also stillage from bioethanol production) capacities showed significant growth in the past years, despite the use of biomethane in different sectors.

7.4 Options for Flexible Production of Liquid and Gaseous Biofuels

Regarding the general options for flexible operation in terms of demand-oriented biofuel supply, biofuel production plants are not comparable to those used for electricity and/or heat/cooling. They can usually be compared with conventional chemical process engineering facilities. Such facilities are usually either running on nominal load mode or not; the part load mode typically used for power production by applications such as combined heat and power engines are not usual for plants producing biofuel. This is due to the fact that products like biofuels can usually be stored much easier than e.g. electricity. The reasons for this so called static operation include relatively easy operation and controlling. Furthermore, most of the facility units are most efficient when operated at their designed nominal load.

Since the most inflexible step has an effect on the general system flexibility, intermediate storage, raw materials and various products are utilized in order to increase the system flexibility. In terms of biofuels the possible ways to achieve flexible plant operation concentrate on the following key objectives:

- Flexible management of raw material input or other input streams such as auxiliaries (reaction media, catalysts),
- Flexibility management of plant operation in terms of main and by-products, including provision of products with high flexibility in application.

The mentioned objectives are mainly driven by the respective market situation which is dependent on external disturbances like fluctuations in the resource and product markets (e.g., volatile and dynamic price developments), policy framework and certain subsidies.

Some exemplary approaches for existing and new concepts will be discussed at a later point.

Table 7.1 Typical technical characteristics and status quo of selected biofuel options

Characteristics	Liquid biofuels					Gaseous biofuels		
	Biodiesel (FAME)	Hydrotreated vegetable oils or esters and fatty acids (HVO/HEFA)	Bioethanol		Synthetic biomass-to-liquids (BTL)	Biomethane/ Biogas	Biomethane/ Synthetic Natural Gas (SNG)	Biohydrogen
Raw materials	Vegetable and animal oils and fats (e.g. rape, soya, palm, grease/used cooking oils, algae oils)	cf. biodiesel	Sugar (beets, cane), starch (corn, wheat, rye)	Lignocelluloses (straw, bagasse, wood, switch grass)	Lignocelluloses (diverse, focus wood, straw, residues such as black liquor)	Sugar and starch, organic residues (e.g. biowaste, manure, stillage)	Lignocelluloses (diverse, focus wood, straw)	Lignocelluloses (diverse, focus wood, straw)
Main conversion steps/plant concept	Vegetable oil production (mechanical or solvent extraction), refining, trans-/esterification, biodiesel treatment	Vegetable oil production (mechanical or solvent extraction), refining, hydrotreating, destillation	Treatment, sugar extraction or hydrolysis/ saccharification, C6 fermentation, distillation, final dehydration	Pretreatment (thermal, acid etc.), hydrolysis, saccharification, C6/C5 fermentation, distillation, final dehydration	Mechanical and thermal treatment (e.g. drying, pyrolysis, hydrothermal), gasification, gas treatment, synthesis (e.g. Fischer Tropsch, FT), hydrocracking, distillation, isomerisation	Silaging, hydrolysis (optional), anaerobic digestion, gas treatment and upgrading	Mechanical and thermal treatment (e.g. drying), gasification, gas treatment, synthesis (methanation), gas upgrading	Mechanical and thermal treatment (e.g. drying), gasification, gas cleaning, reforming, gas upgrading

(continued)

Table 7.1 (continued)

Characteristics	Liquid biofuels					Gaseous biofuels		
	Biodiesel (FAME)	Hydrotreated vegetable oils or esters and fatty acids (HVO/HEFA)	Bioethanol		Synthetic biomass-to-liquids (BTL)	Biomethane/ Biogas	Biomethane/ Synthetic Natural Gas (SNG)	Biohydrogen
By-products[a]	Press extraction, glycerine, salt/ fertiliser, fatty acids, other oleochemicals	(Press extraction), propane, gasoline fractions	Sugar: bagasse/ vinasse/starch: gluten, stillage for DDGS (Distiller's Dried Grains with Solubles), fertiliser, biogas/ biomethane, technical CO_2	Lignin or lignin based by-products, pentoses, stillage products such as fertiliser, biogas/ biomethane, technical CO_2	In case of FT, waxes, naphtha, electricity and heat	Digestate (e.g. as fertiliser), electricity	Electricity and heat	Electricity and heat
Status of technical development	Commercial	Commercial	Commercial	Demonstration	Pilot for FT fuels	Commercial	Demonstration	Pilot
Plant capacity[b]	2–350 MW	255–265 MW (150–1,220 MW)	38–450 MW	0.5–5 MW (35–100 MW)	0.8–5 MW (40–300 MW)	0.5–50 MW	1–10 MW (20–200 MW)	0.5–10 MW (5–100 MW)
Efficiency[c]	38–92 %	58–90 %	30–85 %	37–77 %	49–60 %	53–76 %	62–74 %	47–78 %

<table>
<tr><td>Installed capacity | production worldwide[d]</td><td>50 mn t a^{-1} | 17 mn t a^{-1}</td><td>2.2 mn t a^{-1} | unknown</td><td>90 mn t a^{-1} | 70 mn t a^{-1}</td><td>0.108 mn t a^{-1} | unknown, often only test campaigns</td><td>0.033 mn t/a | unknown, often only test campaigns</td><td>1.21 mn t a^{-1} | unknown</td><td>Not used outside Europe</td><td>Unknown</td></tr>
<tr><td>Installed capacity | production EU[d]</td><td>22 mn t a^{-1} | 8.8 mn t a^{-1}</td><td>1.2 mn t a^{-1} | 840 mn t a^{-1}</td><td>5.8 mn t a^{-1} | 5.2 mn t a^{-1}</td><td>0.019 mn t a^{-1} | unknown, often only test campaigns</td><td>No plants in operation</td><td>0.71 mn t a^{-1} | 0.69 mn t a^{-1}</td><td>0.092 mn t a^{-1} | unknown often only test campaigns</td><td>Unknown</td></tr>
<tr><td>Installed capacity | production Germany[e]</td><td>4.5 mn t a^{-1} | 2,4 mn t a^{-1}</td><td>No plants in operation</td><td>0.9 mn t a^{-1} | 0.6 mn t a^{-1}</td><td>0.001 mn t a^{-1} | unknown, only test campaigns</td><td>No plants in operation</td><td>0.45 mn t a^{-1}| 0.39 mn t a^{-1}</td><td>No plants in operation</td><td>No plants realised</td></tr>
<tr><td>Link to biorefinery platforms[f]</td><td>Oilseed or vegetable oil biorefinery</td><td>Oilseed or vegetable oil biorefinery</td><td>Sugar and starch biorefinery</td><td>Lignocellulosic biorefinery</td><td>Lignocellulosic or synthesis gas biorefinery</td><td>Biogas biorefinery</td><td>Lignocellulosic or synthesis gas biorefinery</td><td>Lignocellulosic or synthesis gas biorefinery</td></tr>
</table>

[a]Usually depending on process design
[b]Related to biofuel output – w/o brackets for current capacities, expected capacities in future in brackets
[c]Overall energetic efficiency according to [18] (ratio of output energy flow of main and all by-products incl. excess process energy to input energy flow of raw material, auxiliaries, external process energy)
[d]Values for 2011 or 2012 based on [1, 22]
[e]Values for 2012 based on [5]
[f]According to Joint European Biorefinery Vision for 2030 [16] and/or the German Biorefineries Roadmap [9]

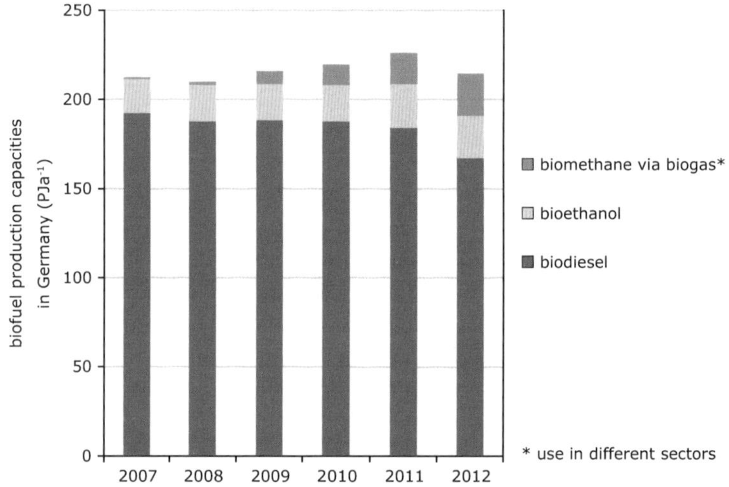

Fig. 7.2 Development of biofuel production capacities in Germany (Adapted from [21])

7.4.1 Approaches for Existing Concepts

As mentioned above existing, biofuel concepts are usually conceived for static operation. Due to the possible storage methods (over a certain period) of the different raw materials, namely liquid fuels in tanks and gaseous fuels like biomethane via the natural gas grid, they are usually running on nominal load and have to deal with production downtimes in case of e.g. volatile market prices of raw materials and product sales challenges. Published information on operation modes is scarce. For biochemical fermentation processes such as for bioethanol and biogas, changes in fermentation can take up to several days, whereas modifications in the running of process engineering plants can take minutes to hours. Examples for biodiesel and bioethanol are given in the following. Flexibility of biomethane from upgraded biogas is discussed in Sect. 8.4.2.

Example Biodiesel Despite the biodiesel production technology (continuous, batch or semi-batch as well as single or multi-feedstock) usually plants run batch-wise on different raw materials (Table 7.1). On the background of the current policy frame in Europe/Germany (which is doubly important for biofuels based on residues for the biofuel quota) plant operators, producing biodiesel based on multi-feedstock technologies and using cooking oil and animal fats, have, for instance, announced an increase of the plant utilization rate of approx. 53–81 % in recent years. This increase is a result of a change in raw material with little production downtime [23]. This does not apply for biodiesel plant operators who use vegetable oils (Table 7.1); in these cases, the rate of plant use decreases from more than 80 % to less than 40 %. The installed overcapacity of biodiesel plants especially in Europe is another reason for this occurrence [12].

Example Bioethanol In regards to flexible plant management, a prominent example is bioethanol production in Brazil. Traditionally, bioethanol is accrued as a by-product of sugar production as the sucrose content of the sugar cane is used in an optimised approach [10]. A number of factors influence the economics of bioethanol production in Brazil, including (i) the development of world prices for sugar, (ii) harvesting results and the quality of sugar cane production, (iii) government-controlled domestic prices for gasoline, (iv) tax policies and (v) exchange rate of Brazilian currency. As Brazil and India are the world's largest producers of sugar they have a major impact on sugar prices. This effects Brazils facility operators in determining how much of its sugar cane production should be refined as sugar or processed to bioethanol [10, 24].

A similar situation occurred in 2007, when a German bioethanol plant operator, who was only producing bioethanol, shut down his plant and sold his contracted raw material, cereal, to the market, which was more profitable than producing bioethanol.

The influence of production plant design (e.g., often efficiency-driven approach) can be illustrated for example by the collapse of the largest corn ethanol biofuel company in the US during the period of high raw material prices around 2007/2008. This operator was using more efficient dry-mill technology (i.e. higher ethanol yield per corn input and lower capital investments). However, due to the limited flexibility of the raw material in question (here just corn grain) and the production of just one primary product (bioethanol), a fair profit margin could not be maintained because of fluctuating market conditions. In comparison, a traditional less efficient wet-mill plant (i.e. lower ethanol yield per corn input and higher capital investments) has a more diverse and adjustable product portfolio (e.g., corn syrup, starch, and ethanol) and thus a better chance of survival in volatile markets [2].

7.4.2 Approaches for New Concepts

While existing biofuel plant concepts are not that flexible it is suggested that in future biofuel and/or biorefinery concepts, operational flexibility needs to be a key issue in order to increase long term economic performance and in effect increase chances of survival when faced with external disturbance [15]. So called flexible polyproduct or polygeneration plants try to produce the most profitable products by altering production according to market fluctuations and thus have the potential to achieve better economic performance compared to conventional static plant operation. However, such flexibility alters the production rate of certain products by oversizing equipment and thus higher capital investments. One of the major challenges therefore is to design polyproduct concepts which take into account the optimal trade-off between operational flexibility and capital cost [2]. Moreover, also plant size of such biorefineries is of major importance with regard to raw material availability and logistic requirements. Especially compared to conventional fossil fuel based refineries or chemical plant they range in the small to medium size.

Example Lignocellulosic Bioethanol The known concepts, which are still in the pilot or demonstration phase for the production of bioethanol based on lignocellulosic biomass (Table 7.1), focus primarily on the production of bioethanol as biofuel [1]. Despite this primary focus, biorefinery concepts also consider the production of bioethanol and other products such as ethylene or carbon acids (e.g. Bioeconomy cluster Leuna in Germany [6]).

Example Biomethane via Bio-SNG Despite the fact that biomethane can be stored for a long period of time in storage facilities in the natural gas grid (see Sect. 8.4.2), the gasifier employed in the process chain is of very limited flexibility (increasing in the order fluidised bed and entrained flow gasifier) and thus also the flexibility of the applied biomass raw material. Catalytic synthesis plants are at present rarely operated in part load mode. However, increasing flexibility in this case is an important research topic. The deployment time of methanation synthesis is approx. 5 min, the cold start a matter of hours, the energy requirement for standby is about 1 % of max capacity [7, 11].

Example BTL/Fischer-Tropsch Fuels In general, the Fischer-Tropsch (FT) process has two important weaknesses: (i) a low overall efficiency and (ii) the production of a wide range of different aliphatic hydrocarbons which makes intensive product separation and treatment necessary for the production of applicable fuels. There are many factors involved when considering the flexibility of the FT process. The produced liquid biofuel can be stored for a long, even indefinite period of time. Different storage technologies such as tanks or storage caverns are well-known for the storage of crude oil and refinery products.

When considering conversion technologies one drawback of FT synthesis as a part of polyproduct refineries is evident: a fixed production rate must be achieved because of the rigorous operational requirements of the gasifier. Thus the overall concept cannot easily be adapted to fluctuating demands [2]. This is described in detail in Chap. 8.

The third aspect is the flexibility of synthesis. Operating the reactor in partial load mode can influence the composition of the aliphatic hydrocarbon mixture because of the changing resistance time. Another important condition is a constant and homogeny temperature profile; this is the second important limiting factor. It is common to operate a plant in full load mode or to stop production completely for an extended period of time. Partial load operation of Fischer-Tropsch synthesis plants is much more complicated than of Bio-SNG plants and is difficult to realize.

One approach that has been investigated is the flexible integrated gasification polygeneration concept, which involves the use of different raw materials (e.g., coal and biomass) and the coproduction of hydrogen, FT fuels as well as methanol, urea and electricity. This approach aims at producing electricity during peak hours while switching to chemicals and fuels during off-peak hours. A high degree of flexibility can be achieved by limiting the operational load of 40–100 % in order to avoid problems in operation. While a complete switch from chemical to electricity production is possible for methanol and urea, for FT fuels the load is restricted to minimum of 60 % in order to avoid a gas turbine load of below 40 % [17].

Example Hydrogen Integration As a lot of the new concepts such as biomethane via Bio-SNG and BTL/Fischer-Tropsch fuels are based on synthesis, applying synthesis gas in addition to the limited flexibility of gasifier for the production of synthesis gas the use of renewable but not biogenous hydrogen is also an option to increase flexibility. Concepts involving the production of additional hydrogen through excess electricity are discussed [8, 7]. The concept behind this follows the ongoing debate surrounding the implementation of intermittent energy sources (IES, e.g. from wind and solar power production) in the existing energy system via so called power-to-gas (PTG) or power-to-liquid (PTL) applications. After this stage, excess electricity from the IES is used for hydrogen electrolysis [11]. In addition to other applications (e.g. accommodation to gas grid, direct use in different industries or for mobility or storage), this hydrogen can be implemented into syntheses like methanation or Fischer-Tropsch (Table 7.1) in order to increase the overall efficiency and economic viability of such SNG or BTL concepts. The addition of hydrogen from electrolysis is one way of adjusting the hydrogen to carbon monoxide ratio. The electrolysis can replace or supplement CO-shift. Furthermore, synthesis through the combined application of hydrogen from electrolysis and carbon monoxide is also possible. However, this is not a biomass application in the narrow sense.

7.5 General Economic and Environmental Aspects

For the efficient realization of these considered concepts, costs and selected environmental aspects are crucial. However, in spite of the fact that several investigations for static process operation have been published, information on flexible biofuel production plant operation is scarce. For this reason only a general overview follows in the section.

Costs Evaluating different cost alternatives is done to identify relative advantages, to compare different options and to determine important influencing factors. Local conditions are relevant in this evaluation. Sensitivity analyses for different biofuels show that in addition to annual full-load hours of the biofuel production plant, raw material costs and total capital investments are of great importance [20]. Furthermore, it should be noted that often market values for raw materials and by-products correlate with each other (e.g., oil seeds and press extraction, starch raw materials and DDGS, Table 7.1) [18]. For example, existing biodiesel production operations have been established with low TCI due to their comparably simple technical complexity. As a result, the impact of annual full-load hours per year is lower. However, the impact of raw material costs is crucial. This is in spite of the fact that there is an increasing tendency to increase total capital investments for biomethane and biofuels based on lignocelluloses in comparison to conventional biofuels. This is often due to more complex technologies and plant designs. However, for future biofuel concepts such as bioethanol, SNG or Fischer-Tropsch fuels, it can be assumed that with regard to biofuel production costs, considerable cost reductions are possible if proposed technical developments are realized [19].

Greenhouse Gas Emissions In regards to the existing frame conditions (e.g., Renewable Energy Directive 2009/28/EC [3] and Fuel Quality Directive 2009/30/EC [4] in Europe), the greenhouse gas mitigation potential of biofuels compared to fossil fuels has become an important value for biofuel marketing and sales. Greenhouse gas emissions are usually determined via life cycle analysis (LCA) which are carried out under different assumptions making it very difficult to compare the results from different studies. For instance, the GHG mitigation potentials for palm oil based biodiesel can range between 36 % and 71 % or 33 % and 66 % when rapeseed is used. The most important drivers for greenhouse gas emissions are (i) biomass production and (ii) biomass conversion to biofuel, including the overall efficiency of the designed concept [18]. These drivers are also important for the achievement of more flexible plant operation.

7.6 Conclusion

Through the consideration of a demand-orientated supply of biofuel for the transport sector, whilst also taking into account the most relevant technologies and concepts for the production and supply of the most important liquid and gaseous biofuels and their current status, the following can be concluded:

- Biofuel concepts are usually unique. They are dependent on the specific local conditions and infrastructure, the equipment provider and often the level of optimization, which is determined by the biofuel production plant operator.
- Flexible part load operation of biofuel plants is not common in comparison to most other process engineering plants in the chemical industry. Flexible operation for fuel synthesis processes is currently a research topic and is not ready for implementation. Today, biofuel plants are usually established as multiproduct plants such as biorefineries. Intermediate storage, raw materials or various products can also be used to increase system flexibility of such biofuel systems, especially when taking into account that the most inflexible step affects the system flexibility.
- Flexible management (i) of raw material and other input streams like auxiliaries (reaction media, catalysts) and (ii) of plant operation in terms of main and by-products, including the provision of products with high flexibility in application, is much more common than part load operation.
- Future flexible polyproduct or polygeneration plants will try to produce the most profitable products altering production according to market fluctuations. These plants will also have to face the major challenges of designing polyproduct concepts, which take into account the optimal trade-off between operational flexibility and capital cost.
- There are almost no investigations of flexible plant operation which consider costs and environmental issues. However, the most important drivers are raw materials (supply costs and emissions related to their production and supply) as well as conversion to biofuels (plant efficiency and annual load).

References

1. D. Bacovsky, N. Ludwiczek, M. Ognissanto, M. Wörgetter, Status of Advanced Biofuels Demonstration Facilities in 2012. A report to the IEA Bioenergy Task 39, T39-P1b (2013)
2. Y. Chen, Optimal Design and Operation of Energy Polygeneration Systems. Ph.D. Thesis, Massachusetts Institute of Technology, 2013. http://yoric.mit.edu/sites/default/files/ChenThesis.pdf
3. European Union, Directive 2009/28/EC of the European Parliament and of the Council of 23 April 2009 on the promotion of the use of energy from renewable sources and amending and subsequently repealing Directives 2001/77/EC and 2003/30/EC, Official Journal of the European Union (2009)
4. European Union, Directive 2009/30/EC of the European Parliament and of the Council of 23 April 2009 amending Directive 98/70/EC as regards the specification of petrol, diesel and gas-oil and introducing a mechanism to monitor and reduce greenhouse gas emissions and amending Council Directive 1999/32/EC as regards the specification of fuel used by inland waterway vessels and repealing Directive 93/12/EEC, Official Journal of the European Union (2009)
5. F.O. Lichts, World Ethanol & Biofuels Report Bd. (2008–2013)
6. Th. Gabrielczyk, "Wir wollen die maximale Wertschöpfung" – Interview mit Prof. Dr. Thomas Hirth im "transcript" (2012). http://www.bioeconomy.de/wir-wollen-die-maximale-wertschoepfung-interview-mit-prof-dr-thomas-hirth-im-transcript/. Accessed 9 2013
7. M. Gassner, F. Maréchal, Thermo-economic optimisation of the polygeneration of Synthetic Natural Gas (SNG), power and heat from lignocellulosic biomass by gasification and methanation. Energy Environ. Sci. **5**, 5768–5789 (2012). doi:10.1039/c1ee02867g. http://infoscience.epfl.ch/record/174475/files/c1ee02867g_ownlayout.pdf
8. M. Grasser, F. Maréchal, Thermo-economic optimization of the integration of electrolysis in synthetic natural gas production from wood. Energy **33**(2008), 189–198 (2008)
9. German Federal Government, *Biorefineries Roadmap as Part of the German Federal Government Action Plans for the Material and Energetic Utilisation of Renewable Raw Materials* (Berlin, 2012)
10. J. Giersdorf, *Politics and Economics of Ethanol and Biodiesel Production and Consumption in Brazil*. DBFZ Report No 15 (Leipzig, 2013). ISSN 2190-7943
11. L. Grond, P. Schulze, J. Holstein, *Systems Analyses Power to Gas: A Technology Review*. Part of TKI project TKIG01038 (KEMA Nederland B.V., Groningen, 2013), http://www.dnv.com/binaries/dnv%20kema%20%282013%29%20-%20systems%20analyses%20power%20to%20gas%20-%20technology%20review_tcm4-567461.pdf
12. IEA, *Technology Roadmaps – Biofuels for Transport* (International Energy Agency, Paris, 2011)
13. IEA, *World Energy Outlook 2011* (International Energy Agency, Paris, 2011)
14. M. Kaltschmitt, H. Hartmann, H. Hofbauer (eds.), *Energie aus Biomasse: Grundlagen, Techniken und Verfahren*, 2nd edn. (Springer, Berlin/Heidelberg, 2009)
15. N. Kou, Biofuel production system with operation flexibility: Evaluation of economic and environmental performance under external disturbance, Purdue University, 2011, http://docs.lib.purdue.edu/dissertations/AAI3477687/
16. C. Luguel et al., Joint European Biorefinery Vision for 2030. Star-colibri Strategic Targets for 2020 – Collaboration Initiative on Biorefineries, Brussels, 2011
17. J.C. Meermann, A. Ramírez, W.C. Turkenberg, A.P.C. Faaij, Performance of simulated flexible integrated gasification polygeneration facilities. Part A: A technical-energetic assessment. Renew. Sustain. Energy Rev. **15**(6), 2563–2587 (2011)
18. F. Müller-Langer, A. Gröngröft, S. Majer, S. O'Keeffe, M. Klemm, Options of biofuel production – status and perspectives, in *Transition to Renewable Energy Systems*, ed. by D. Stolten, V. Scherer (Wiley, Weinheim, 2013). ISBN 987-3-527-33239-7

19. F. Müller-Langer, M. Kaltschmitt, Biofuels of tomorrow – concepts and their assessment, in *Conventional and Future Energy for Automobiles: [Proceedings 2013]/9th International Colloquium Fuels, 15–17 Jan 2013*, ed. by J. Bartz Wilfried (TAE, Ostfildern, 2013)
20. F. Müller-Langer, S. Majer, A. Perimenis, Biofuels – a technical, economic and environ-mental comparison, in *Encyclopedia of Sustainability Science and Technology*, ed. by R.A. Meyers (Springer, New York, 2012). ISBN 978-0-387-89469-0
21. K. Naumann, K. Oehmichen, M. Zeymer, Monitoring Biokraftstoffsektor. DBFZ Report 11, 2. Auflage (2014). ISSN 2190-7943
22. K. Naumann, K. Oehmichen, M. Zeymer, F. Müller-Langer, M. Scheftelowitz, P. Adler, K. Meisel, M. Seiffert, *Monitoring Biokraftstoffsektor* (Deutsches Biomasseforschungszentrum (DBFZ), Leipzig, 2012)
23. Petrotec, Petrotec – A leader in waste to energy production. 7th annual general meeting, Düsseldorf, 29 May 2013, http://www.petrotec.de/core/cms/upload/pdf/hv2013/AGM_Presentation_FY2012_final.pdf. Accessed 09 2013
24. R. Wisner, Brazil Ethanol Developments & Implications for the U.S. Ethanol Industry. AgMRC Renewable Energy & Climate Change Newsletter, Oct 2012, http://www.agmrc.org/renewable_energy/ethanol/brazil-ethanol-developments--implications-for-the-us-ethanol-industry/#. Accessed 09 2013

Chapter 8
Intermediate Biofuels to Support a Flexible Application of Biomass

Eric Billig, Janet Witt, Marco Klemm, Claudia Kirsten, Jan Khalsa, and Daniela Thrän

Abstract As the previous book chapters concluded, the future bioenergy provision concepts for power, heat and transport fuels are characterised by more complex demands. A future energy market is characterised by the need for a sustainable flexible energy carrier with homogeneous properties for application in the fields of CHP, heat and fuel. To some extent these energy carriers are already available today (see Chaps. 4, 5, 6 and 7). However, in many cases untreated biomass cannot fulfil the requirements of existing and future conversion processes or demands respectively. As far as solid biofuels are concerned, the high moisture content of untreated biofuels coupled with a low energy density and high biological activity require the development of often costly storage, transport and conversion techniques. Various research activities are still ongoing to improve the utilisation of biofuels in existing and future technologies, available infrastructure and therefore also in logistic and storage issues. A similar development can be observed regarding the biogenic substitutes for natural gas (biomethane, bio-SNG). Such upgraded "new" – or rather "advanced" – solid and gaseous biofuels are high energy value products for gasification and combustion in industrial conversion plants as well as for domestic applications with excellent advantages in flexible energy provision. The amount of advanced solid biofuels in the markets of heating and power or combined heat and power systems will increase, as will the share of the biogenic substitutes for natural gas with further development and process optimisation.

E. Billig (✉) • J. Witt • M. Klemm • C. Kirsten • J. Khalsa
Deutsches Biomasseforschungszentrum GmbH – DBFZ,
Torgauer Str. 116, 04347 Leipzig, Germany
e-mail: eric.billig@dbfz.de; janet.witt@dbfz.de; marco.klemm@dbfz.de; Claudia.Kirsten@dbfz.de; Jan-Hari-Arti.Khalsa@dbfz.de

D. Thrän
Department of Bioenergy, Helmholtz Centre for Environmental Research – UFZ, Permoset Straße 15, 04318 Leipzig, Germany

Deutsches Biomasseforschungszentrum – DBFZ, Torgauer Straße 116, 04347 Leipzig, Germany

Bioenergy Systems, University of Leipzig, Grimmaische Straße 12, 04109 Leipzig, Germany
e-mail: daniela.thraen@ufz.de

D. Thrän (ed.), *Smart Bioenergy*, DOI 10.1007/978-3-319-16193-8_8

This chapter reviews the current developments in selected biomass pretreatment processes and their intermediate biofuel products that have the potential to increase flexible bioenergy production in the short and mid-term. On the one hand, these include biomass densification without thermal treatment as well as torrefaction and hydrothermal treatment for producing intermediate solid biomass. On the other hand, technologies for biogenic substitutes for natural gas are evaluated. The focus lies on the surplus value of the technologies in terms of flexibility during energy production or use of the advanced solid biofuels or biogenic substitutes for natural gas as intermediate bioenergy carriers.

8.1 Introduction

A future energy market is characterised by the need for sustainable energy carriers that fulfil the demands of the more flexible application that is anticipated in the fields of combined heat and power (CHP), heat and fuel. These future energy carriers will feature homogeneous properties for flexible provision, fast reaction times when operating during the conversion process, usage in small, medium and large scale plants as well as the utilization of a broad biomass resource base.

In the future, a sharp distinction between the fields of CHP, heat and fuel will no longer be possible and interactions will be commonplace, (see Chap. 2). As far as biomass is concerned, two types of future or intermediate fuels are promising (see Fig. 8.1). These are the advanced solid biofuels and biogenic substitutes for natural gas.

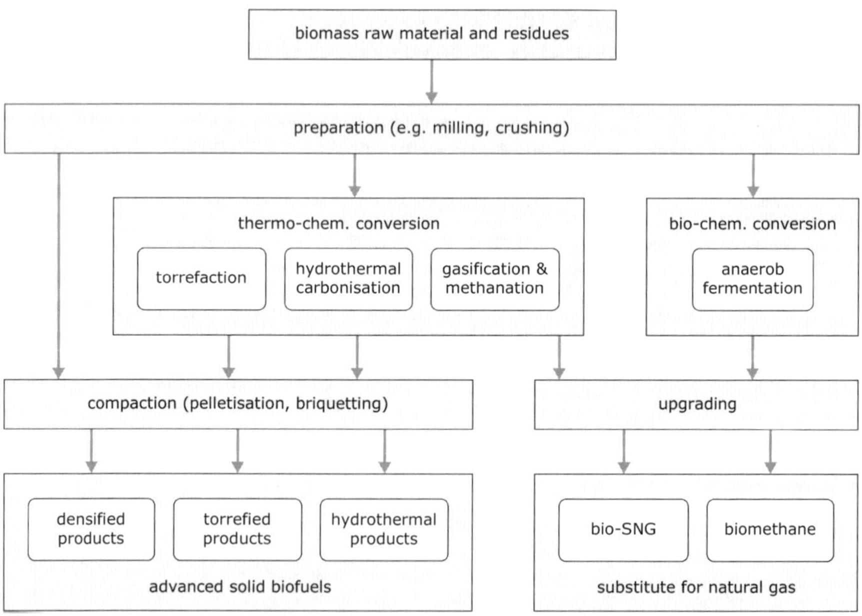

Fig. 8.1 Selected intermediate bioenergy carriers for supporting the flexible application of biomass

From the advanced solid biofuels, the densified and thermally treated ones (torrefaction and hydrothermal carbonisation) are considered and from the biogenic substitutes for natural gas, biomethane or bio-SNG are considered. Each biofuel has its own advantages in terms of flexible properties compared to conventional fuels. To some extent, these energy carries are already available today or still under development.

The potential contribution of the advanced solid biofuels and the gaseous biofuels to a future energy supply system is different:

Advanced solid biofuels have advantages over loose material (wood chips, straw chops, mixtures of landscape residues etc.) in combustion and gasification systems due to improved product qualities of the fuels, which can be summarized as follows:

- Homogeneous material (carbon content, heating value, water content, shape and form, higher mechanical durability, higher bulk density, etc.)
- Reduced water retention leading to no or only low biological degradation caused by a low moisture content, reduced self-heating risk and increased storage stability
- Low dust formation during the biofuel transportation and redistribution, reduced health and safety risks (e.g. dust explosion), improved handling of logistics
- Fuel quality committed to improving the desired conversion characteristics (e.g. avoiding emissions, slagging, corrosion problems) from blending with additives, several other raw biomass materials or thermal treatment

Biomethane extend the possibilities for the application of biomass in all established applications of natural gas, such as efficient and flexible processes for power generation, transport fuel, chemical base materials for further synthesis, easy controllable plants and innovative application technologies such as fuel cells and better storage opportunities. With the existing natural gas grid, biomethane can be transported and stored in the existing infrastructure without additional investment costs. Furthermore, a wide range of feedstocks can be converted.

Advanced solid biofuels as well as biomethane, play an important role due to their properties in supporting a modern energy system in the fields of heat, power and fuel and are therefore worth looking at more closely. Nevertheless, there are many other biogenic fuels, such as liquid fuels for transport, which are also relevant and should therefore also be considered (see Chap. 7).

8.2 Advanced Solid Biofuels

Intermediate solid biofuels can be divided into the densified biofuels such as standard pellets or briquettes and the thermally treated biomass which can also be in the form of pellets or briquettes but produced by a different substrate and/or process. Thermally-treated products are not ready for the market at present, but they aim to come onto the market as a commodity fuel in the short to mid term and contribute to a more flexible energy provision. Different thermal-treatment processes can be used to produce advanced solid biofuels i.e. torrefaction, hydrothermal carbonisation, steam explosion or fast pyrolysis, whereby the first two options are presented later on, because these are the most developed ones.

8.2.1 *Densified Solid Biofuels*

Pressure agglomeration processes such as briquetting and pelletizing are used to improve the mechanical and physical properties of solid biofuels. The aim is to convert particles or fibers into products with reduced volume and designated forms and properties such as reduced moisture content. These special fuel properties are especially suitable for long distance transport, efficient storage requirements as well as advantages in process control of the conversion system through the automatic feeding of a fuel with homogenous fuel properties. These properties are essential for a flexible usage.

Raw Material

Generally, all kinds of solid biomass can be densified if a feedstock-specific process adaptation can be assumed. The densification of woody materials is particularly relevant, because the energy density of woody biomass is naturally high compared to other solid raw materials [16]. Predominantly low-grade wood fractions such as wood residues and the by-products of saw mills and the wood processing industry are used, due to their low moisture content (reduced drying demand) and relatively homogeneous material properties. However, as the biomass potential is limited, there is an increasing interest in alternative green wood fractions, such as forest residues, stem wood (from catastrophic events such as storms, windthrow or bark beetle infestations) or short rotation coppice [33]. Prospectively, the use of green wood in pellet production is expected to increase [4] as is the use of herbaceous materials such as straw and hay.

Process

The densification of solid biomass involves drying down to a moisture content of 15–20 %, milling and conditioning the material in the form of regulating the water content and improving product quality e.g. adding binders to improve durability, densification and cooling [16]. To achieve the required standardised physical-mechanical properties of the end product, an optimal parameter combination is required. The production of high-quality fuel pellets or briquettes is very similar, with the difference being in the product size (pellets have a diameter less than 25 mm, briquettes are larger [9]). Moreover, the briquetting process doesn't normally require cooling or sieving. Figure 8.2 illustrates the process.

More information about the properties of standard pellets can be found in Table 8.1, where they are compared with torrefied pellets, wood chips and coal.

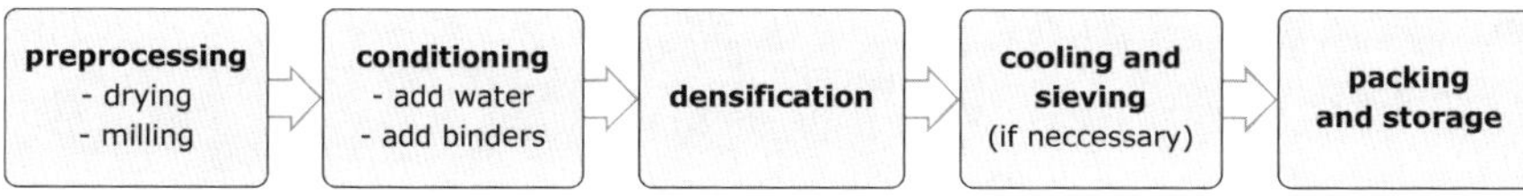

Fig. 8.2 Process steps for densified solid biofuels

Table 8.1 Comparison of standard and torrefied pellets with wood chips/coal (Adapted [34])

	Wood chips	Wood pellets	Torrefied wood pellets	Coal
Moisture content (wt%)	30–55	7–10	1–5	10–15
Calorific value (LHV, MJ/kg) as received	7–12	15–17	18–24	23–28
Volatile matter (wt%, dry basis)	70–84	75–84	55–80	15–30
Fixed carbon (wt%, dry basis)	16–25	16–25	22–35	50–55
Bulk density (kg/l)	0.20–0.30	0.55–0.65	0.65–0.80	0.80–0.85
Vol. energy density (GJ/m^3)	1.4–3.6	8–11	12–19	18–24
Hygroscopic properties	Hydrophilic	Hydrophilic	(Moderately) hydrophobic	Hydrophobic
Biological degradation	Fast	Fast	Slow	None
Milling requirements	Special	Special	Standard–feedstock-specific	Standard
Product consistency	Limited	High	High	High

wt% = weight percentage

Energy Balance

Depending on the quality of the feedstock, about 3–10 % of the energy content of the biomass is necessary for the production of pellets. The specific energy consumption of a wood pellet press is between 1.3 % and 2.7 % based on the energy content of the pellets [33]. In the case of wood briquette production this percentage may be lower. However, the energy balance of the process – which directly influences the cost balance – depends mainly on the raw material used.

Cost Range

The largest cost factors in pellet production are the raw material (43–73 %) itself and the potential drying need (ca. 35 %) [25]. When compared with the densification of woody material, the production costs of alternative fuel pellets (e. g. hay or straw pellets) are expected to be slightly lower due to a reduced demand for drying.

Stage of Development

Pellet mills and briquette presses are state of the art and available on the market. The international product standard (EN 14961-2/3; ISO 17225-2/3) has supported the implementation of wood pellets and briquettes as a commodity fuel on the market for almost 20 years. In 2012, the European wood pellet market became the world's largest market with a production of 10.5 million tons. Between 60 % and 70 % of the world's market volume of 22.4–24.5 million tons were consumed in the EU in 2012. The four largest pellet-producing countries in the EU are Germany (2.2 million t), Sweden (1.2 million t), Latvia (1 million t) and Austria (0.9 million t). The largest wood pellet exporters to the EU are the USA with 1.8 million t and Canada with 1.3 million t. Russia, the Ukraine and Belarus follow and the mid -term expectations for a future growth in the market are promising with an estimated demand that is triple to tenfold [10].

8.2.2 Torrefied Fuels

Torrefaction is a thermochemical pretreatment for carbonaceous feedstock, including a multitude of different biomasses [32]. Torrefied biofuels that are densified show almost the same properties as densified biofuels without thermal treatment. When compared with standard pellets however, the pellets form torrefied biomass show better properties for grinding, chemical and biological degradation during storage and they are expected to show improved combustion or gasification behaviour.

However, it is important to consider the high reactivity of ground torrefied biomass during storage, which calls for inert conditions if spontaneous combustion is to be prevented [29].

Raw Material

Different kinds of dry feedstock can be used for the torrefaction of biomass. Currently, the research focus is on woody biomass but straw and other biomass residues are gaining more popularity. The physio-chemical composition of the product will largely affect how much of the raw materials will be transformed into a gaseous phase or remain as a solid [5]. So far torrefaction has mainly only been performed on woody biomass.

Process

The torrefaction process can best be described as a mild pyrolysis under inert or almost inert (small percentage of oxygen) conditions and near atmospheric pressure within the temperature range of 200–320 °C. By increasing the torrefaction temperature, the amount of volatiles released from the biomass increases while hemicellulose, lignin and cellulose are decomposed. The torrefied product is influenced by the biomass composition, the heating rate and the residence time. The degree of torrefaction is often described by a combined mass- and energy yield while the relative reduction in volatiles can also be used [26].

In Table 8.1 key parameters of torrefied and standard pellets are compared with wood chips and coal.

Energy Balance

Torrefaction requires certain energy input which can – in the best case – be provided by an auto-thermal operation of the process. Therefore, it is important to capture and utilise as much energy as possible that is contained in the torrefaction gas and to recycle it to the torrefaction process and drying of the biomass prior to torrefaction. The energy that is transferred into the torrefaction gas, typically around 10 % (Fig. 8.3) is strongly affected by the torrefaction temperature and residence time. These two are therefore the key parameters that will affect the energy efficiency of

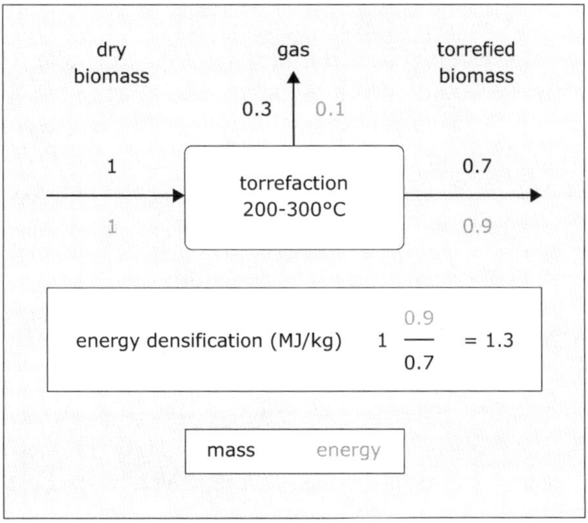

Fig. 8.3 Energy/mass balance of the biomass torrefaction process [17]

the overall process. Only when the energy content of the torrefaction gas is large enough to balance the heat demand of drying and torrefaction, can an auto-thermal operation be achieved [29].

Additionally, the densification of torrefied material is more energy-intensive than palletisation/briquetting of untreated biomass due to lower self-binding forces in the material (reduced hemicellulose and lignin). On the other hand, less energy is required to grind torrefied materials, particularly enhancing energy efficiency and enabling its utilisation in dust boilers.

Cost Range

Generally, it is expected that additional thermal treatment processes of the biomass automatically result in higher production costs than the densification of untreated material. At present the prices for torrefied (and densified) biofuels range between three to tenfold of the price of standard wood pellets, as described in Sect. 8.2.1. The cost variation highly depends on factors, such as raw material availability and quality, treatment technology, logistics and end-use requirements. Furthermore, the technology has not yet been made commercially available on the market, which also contributes to higher costs compared to standard pellets. Optimistic market observers assume that torrefaction will become commercially available within the next 1–2 years [8].

Stage of Development

Numerous activities exist pronouncing a worldwide installation of torrefaction plants with a total production capacity between 300,000–500,000 t, mainly installed in the U.S. or European market [34]. Worldwide, approximately 50 technology developers or initiatives are currently battling on the market to present the first commercially-run torrefaction plant. Different reactor designs for the production of torrefied biomass are still at the pilot or demonstration stage with preliminary demo-units in operation. The most important concepts seem to be the compact moving bed and the fluidised moving bed reactors [8]. However, at the current stage of development, there are only a handful of existing installations operating as demonstration or pilot plants and producing several kilograms to several thousand tons. Fuel standardisation was started back in 2012 [15].

8.2.3 Hydrothermal Carbonised Fuels

In the hydrothermal process, the hydrothermal carbonisation (HTC) is the process for the production of a solid fuel, so-called "biocoal" from a wet feedstock. The process is performed under high temperatures and pressures with a wide range of feedstock. The product can be used for energy provision as well as for material use (i.e. for soil improvement) [30].

Raw Material

In addition to biomasses with established applications in the combustion or the biogas processes, there is an important potential of wet and not very biodegradable biomass such as food industry wastes, municipal biowaste, digestates from biogas processes and sewage sludge. The utilisation of this potential is of major importance for extending the feedstock base of the bioenergy supply [32].

Process

The HTC is a hydrothermal process for the production of a solid fuel, the so called "biocoal". It is performed in pressurised hot water at 180–250 °C. The pressure is determined by the temperature because liquid water is necessary as a reaction agent that is why 10–40 bar are common. In some cases, an acidic catalyst is used. Currently, an operation time of 1.5–6 h is standard. Because the hydrothermal carbonisation process (HTC) takes place in liquid water, no preliminary drying is needed. Generally, the process consists of a pre-treatment where the biomass is mixed with water, the conversion is influenced by heat and pressure and a post-processing where the water content of the product is reduced, see Fig. 8.4.

In contrast to biological processes, hydrothermal processes are able to convert all organic fractions including lignin. The properties of biocoal mainly depend on the reaction conditions. With increasing residence time, the product changes its state to become more like coal. Elementary analysis values are listed in Table 8.2 Examples of fuel properties of biocoal from hydrothermal carbonisation process HTC compared with brown coal and biomass (dry matter) [7, 28].

During the conversion of biomass into biocoal the reaction mechanisms of hydrolysis, dehydration, decarboxylation, aromatization and condensation polymerisation are involved.

An efficient removal of the reaction agent water is of major importance for the economic production of an applicable product. Because of the altered structure, water can be removed much easier using mechanical processes compared to the water content of raw biomass. This is one of the major advantages of HTC, enabling an efficient fuel production.

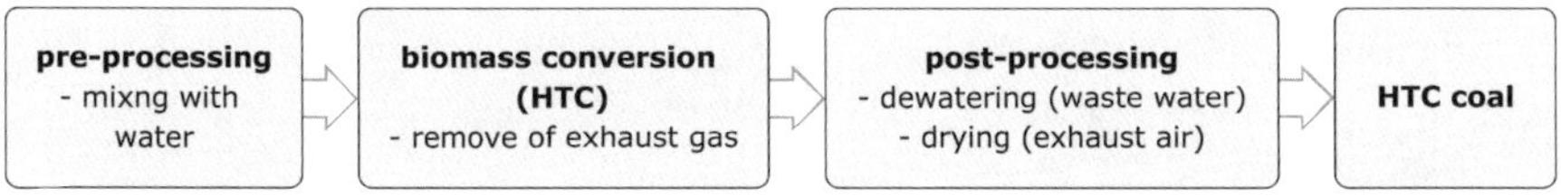

Fig. 8.4 Process steps of hydrothermal carbonisation process

Table 8.2 Examples of fuel properties of biocoal from hydrothermal carbonisation process HTC compared with brown coal and biomass (dry matter) [7, 28]

	HTC-biocoal			
	From green waste	From municipal bio-waste	From digestate (dry fermentation)	Brown coal briquette
Heating value (LHV)	16.7 MJ/kg	19.4 MJ/kg	18.1 MJ/kg	24.9 MJ/kg
Ash content	27 %	17 %	22 %	4.2 %
Sulphur content	0.13 %	0.2 %	0.3 %	0.3 %
Nitrogen content	1.1 %	1.8 %	1.6 %	0.74 %
Chlorine content	0.04 %	0.08 %	0.18 %	0.027 %

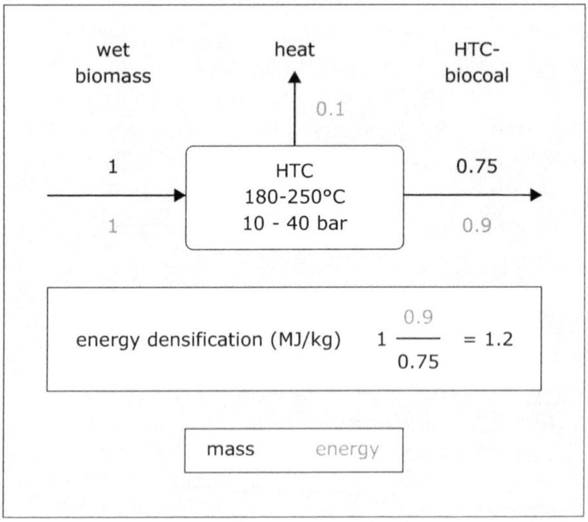

Fig. 8.5 Energy balance for HTC

Energy Balance

The types of feedstock used as well as the plant design strongly influence the energy balance. Figure 8.5 shows an example of an energy balance for the HTC process. It is normally found that the conversion reaction occurs after pre-heating the biomass and the reaction agent water without any need for or surplus of energy. Another important energy demand is the heat for product drying. Mechanical water separation is possible, if a higher dry matter content is needed, then thermal drying is necessary.

Stage of Development

Generally speaking, it can be said that the transformation from the laboratory scale to the technical scale is currently ongoing. Preliminary demonstration plants are in their initial operation phase, for example in Halle (Germany) a unit for the application of landscape management matter, bio-residues and fermentation residues has been installed [20]. Because of the state of development, substantive economic figures cannot currently be published. Compared to torrefied solid biofuels for HTC products, fuel standardisation has not yet been started.

8.3 Biomethane

With the overall goal of this book in mind, the main focus of this subchapter is the synthetic methane from the gasification of biomass (bio-SNG) as well as the biomethane from the biochemical conversion process with an upgrading of biogas to a methane-rich gas with the focus on flexibility.

Bio-based synthetic natural gas (bio-SNG) and biomethane from upgrading biogas are biogenic substitutes for natural gas with methane (CH_4) that is produced as much as 99.9 % pure. Bio-SNG is produced thermo-chemically, primarily from lignin-based substrates such as wood. Biomethane is produced biochemically by cleaning and upgrading raw biogas. The state-of-the-art, further perspectives and the advantages of a flexible usage are outlined in the following.

8.3.1 Bio-based Synthetic Natural Gas Bio-SNG

Raw Material

The SNG-process is a particularly promising alternative for dry and solid biomass with low degradability and high lignin content such as wood and straw. In contrast to the Power-to-Gas process, which utilises carbon dioxide and hydrogen, the SNG-process converts carbon monoxide and hydrogen generated in a biomass gasifier into methane.

Process

A typical SNG-plant incorporates the following process steps: biomass pretreatment, gasification, synthesis gas treatment, methane synthesis and methane separation (see Fig. 8.6). Because of the complex technology involved, SNG-plants are favoured for medium to large scale facilities with up to 500 MW bio-SNG

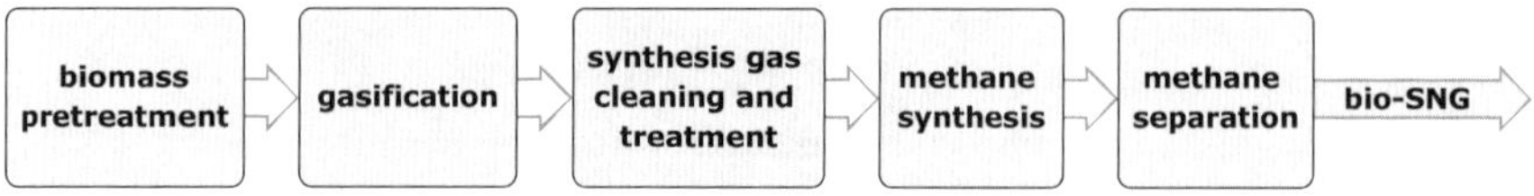

Fig. 8.6 Bio-SNG process chain

output. The two major process steps are gasification and synthesis, although the other steps are also crucial for success.

Gasification is defined as the conversion of a solid or liquid fuel, here biomass, to a gaseous fuel, mainly hydrogen and carbon monoxide, in a reaction with an added reaction agent. The gasification process consists of the typical process steps: drying, pyrolytic decomposition, oxidation, reduction and gas phase reactions. For these reactions, many different reactor concepts are available, including fixed bed, fluidised bed and flow reactor concepts [3, 21, 23, 24, 27].

The choice of gasifier has a major influence on the economic size of a facility, the required biomass pre-treatment, the synthesis gas treatment as well as the plant's flexibility.

- Fluidised bed gasifiers can convert fuel particles with an average size of several millimetres. Because of the chemical equilibrium, the total carbon content cannot be converted. If a single-stage fluidised-bed reactor is applied, then the unconverted carbon will remain in the ash. In a two-stage gasifier such as the Fast Internally Circulating Fluidised Bed Gasifier (FICFB) the remaining carbon is converted in the second chamber for heat supply whereas in the first reactor, gasification will take place. Preliminary demonstration plants are now in operation.
- In an entrained-flow gasifier the reactions take place while the particles are transported by the fluid phase. The chopping of biomass to the necessary particle size (less than 1 mm) is expensive and normally only possible with a thermal pre-treatment, such as torrefaction. Entrained flow gasifiers are mainly suitable for large-scale SNG-plants, where several hundred MW are the norm.

The following **methane synthesis** or **methanation** is the exothermic, catalytic conversion of the synthesis gas to methane (CH_4), carbon dioxide and water. Side reactions are the water-gas-shift-reaction and the Boudouard reaction. Due to the selectivity, activity and costs, commercial projects focus on Ni catalysts, whereas others are possible [22]. Common methanation reactors are adiabatic fixed bed and isothermal fluidised bed reactors. Currently, the bio-SNG production is exclusively demonstrated in combination with steam and/or oxygen-blown gasification because of the high nitrogen content in the synthesis gas when air is used as a reaction agent.

Because of the similarities between SNG and biogas (methane, carbon dioxide and water content) processes similar to biogas upgrading can be employed, see Sect. 8.3.2. Depending on usage and distribution, further steps such as compression for gas grid injection have to be applied.

Stage of Development

The production of regenerative methane from the thermo-chemical pathway is still under development with few demonstration plants so far. An implementation of this technology for SNG production is expected over the next couple of years [13].

8.3.2 Biomethane from Upgrading of Biogas

Raw Material

The basic raw material for biomethane from the biochemical pathway is the same as for biogas production with a combustion purpose. Usually it is sourced from energy crops, agricultural residues e.g. straw, manure and industrial or organic waste. Depending on the basic feedstock and the specific digestion properties, a raw biogas with various gas qualities is produced; see Chap. 2.

In summary, raw biogas is essentially a mixture of methane and carbon dioxide, while the proportion of methane is usually higher than that of carbon dioxide. Because of its chemical composition, raw biogas cannot replace natural gas without further treatment.

Process

To exploit its full potential, the raw biogas has to be upgraded. After upgrading the biogas to biomethane, it shows almost the same properties as natural gas and can therefore be fed into the existing natural gas grid infrastructure. There are various ways of upgrading biomethane from biogas. In essence, they all reduce the CO_2 content while enriching the CH_4 content of the raw biogas. Depending upon the upgrading technology and the raw gas quality, pre- and post-treatment are required. Figure 8.7 shows the schematic process of biogas upgrading.

Pre-treatment Depending on the composition of the raw biogas and the CH_4 enrichment technology, different pre-treatments have to be applied, mainly to reduce sulphur, water or other undesired components such as siloxane.

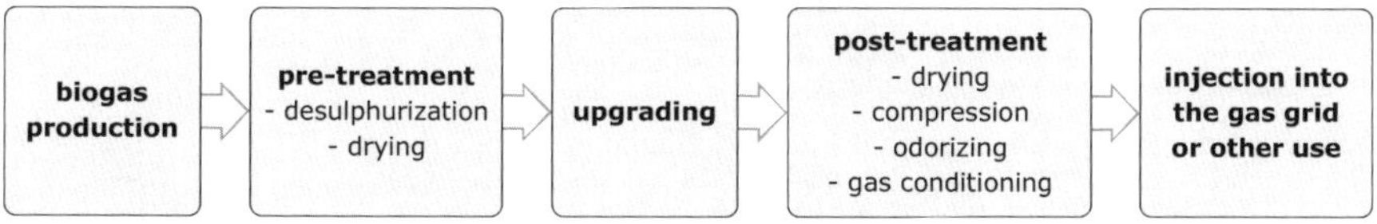

Fig. 8.7 Schematic diagram of the biogas upgrading process

CH_4-Enrichment (Biogas Upgrading) Currently there are five main technologies being used for biogas upgrading. Other technologies such as cryogenic separation, liquefaction and small-scale upgrading processes are still being developed. The upgrading capacity from the state-of-the-art technologies ranges between 250 and 2,800 m^3/h STP (Standard Temperature and Pressure) raw biogas input, see Table 8.3.

- Water scrubbing (WS): dissolve of CO_2 in water under pressure in an absorption column. A methane-rich gas leaves the top of the absorption column while the CO_2 is released in a second column by pressure release.
- Pressure swing adsorption (PSA): based on selective adsorption of CO_2 on adsorbents such as active carbon. The process operates under pressure. For the reuse of adsorbents, the CO_2 desorbs from the adsorbents by pressure release.
- Chemical absorption: dissolving of CO_2 in a solvent (amine) in an absorption column. A methane-rich gas leaves the top of the absorption column while the CO_2 is released in a second column through heating of the solvent.

Table 8.3 Overview[a] of raw biogas CH_4 enrichment technologies [1, 2, 12, 31], (costs for biogas production, upgrading and injection into the gas grid [according to the German version [13]])

Parameter	WS	PSA	Chemical absorption	Physical absorption	Membrane separation
Operating pressure in bar(a)	5–10	4–7	1–3	4–8	5–10
Regenerating temperature in °C	–	–	120–160	70–80	–
Plant size range[b, c] in m^3/h STP	350–2,800	400–2,800	500–2,000	250–2,800	400–700
Electric energy demand[b] in kWh/m^3 STP	0.17–0.23	<0.19	0.09	0.23–0.27	0.24
Thermic energy demand[b] in kWh/m^3 STP	0	0	0.6	Internal provision from lean gas	0
Max. extern usable heat[b] in kWh/m^3 STP	0.06–0.18	<0.1	0.3	0.12–0.13	0.36
Methane slip before lean gas treatment in %	<2	<2	<0.1	1–4	<5
Methane purity[a] in %	95–99	95–99	>99	95–99	95–99
Lean gas treatment necessary?	Yes	Yes	No	Yes	Yes
Specific biomethane costs in €ct/kWh Hs	6.2–8.3	6.4–8.5	7.1–8.1	6.5–8.7	8.3–8.8

[a]Values in operation can differ and can be customised
[b]Referring to raw biogas
[c]Currently available on the market

- Physical absorption: similar to water scrubbing. Instead of water, an organic solvent with a higher absorption capacity of CO_2 is used. The process operates under pressure, the solvent has to be heated for desorption.
- Membrane separation: the process operates under pressure with 1–3 membrane process steps. The separation is based on the different kinetic diameters of the molecules and the resulting permeation through the membrane. Carbon dioxide and other components (e.g. ammonia, oxygen and hydrogen) transport the membrane to the permeate side while methane mainly remains on the retentate side.

Post-treatment According to applicable law and technical instructions, biomethane must fulfil specific requirements for usage as a transport fuel, for gas grid injection or for other usages. This includes drying, compression, odorising, gas conditioning and lean gas treatments. The post treatment depends upon the upgrading technology, the usage (e.g. gas grid injection) and the local applicable law.

Cost Range

The cost range (Table 8.3) for upgrading biogas depends on various factors. These can be input parameters such as substrate and energy costs, technology factors such as energy consumption and methane slip as well as post-treatment factors such as gas conditioning requirements and the pressure stage. Such upgrading costs can range between 6.2 and 8.8 €ct/kWh Hs [12].

Stage of Development

The first biogas to biomethane upgrading plant was implemented in Germany in 2006. Since then a steady increase has been observed, see Sect. 2.5.2. Although there are now (end of 2013) more than 120 plants in operation [6], there is still need of improvement. The newest technology for upgrading is the membrane process. Manufacturers of those upgrading plants that already exist show an on-going commitment to improving the technologies on offer or have started to implement new ones such as membrane upgrading. To summarize, the upgrading from biogas to biomethane is an expanding market with increasing efficiency. In the long-run, it will be those technologies with the best energy- and cost-efficiency combined with the best operating and maintenance properties that will prevail. At this moment in time, the membrane technology looks promising. So far no uniform standards across the European Union or even worldwide for that matter have been implemented, but are still under development. With the implementation of standardized gaseous fuel properties, a further boost for market penetration is estimated [2, 6, 12].

8.4 Contributing to a Flexible Energy Supply

From the previous chapters of the book it can be concluded that a future energy market will be characterised by the need for a sustainable flexible energy carrier with homogeneous properties for application in the fields of combined heat and power generation, heat and fuel. The previous sections of this chapter introduced two kinds of intermediates: advanced solid biofuels and biogenic substitutes for natural gas. Each of them with their specific flexibility properties has the potential to contribute to a well-balanced energy market.

8.4.1 Flexibility Through Solid Fuels

Flexible bioenergy provision from solid biofuels demands (i) smaller conversion units in the heat sector, (ii) higher technical demands (ramp loads, gasification systems) in the electricity and transport sector, (iii) the capability to widen the resource base from wood to other, less homogenous solid biofuels like straw, residues from gardening etc. and (iv) improved time-dependent application due to ideal storage and easy transport factors (energy production on demand). Therefore, the development of intermediates is the counterpart of the development of new, flexible concepts.

Wood pellets, which have been on the market for nearly 20 years now, are a success story in this field. Due to their fuel properties they are suitable for automated stoves and boilers for small houses as well for medium and large scale boilers for e. g. municipal facilities or industrial applications. The availability of this technology has been one of the starting points for discussing the future options of a flexible energy supply in this chapter. The weakness of the conventional pellet technology is that so far only a limited assortment of woody biomass can be used as feedstock.

Torrefied pellets have the potential to provide the desired quality from wood pellets from a wider resource base. Additionally, with thermo-chemical pretreatment, the fuel properties change towards an even lower degree of biodegradability, become easier to grind and show a more stable reaction time for the particles in the conversion process. So far, torrefied pellets have mainly been developed for the option of using biomass in coal-fired power plants, but an examination of additional markets is already underway.[1] From today's perspective it is only possible to state the systemic advantages of those new properties. For example: future heating systems might be designed smaller because process control is easier. In the long term, new

[1] e.g. in the European-FP7-project "SECTOR – Production of Solid Sustainable Energy Carriers from Biomass by Means of Torrefaction (2012–2015) or in the national project financed by the BMWI "FlexiTorr" (Flexibilisation of energy supply in small bioenergy generation plants due to the use of torrefied biomass), 2013–2013.

concepts are also imaginable that grind the fuel before combustion in other devices (i.e. dust boiler).

The development of HTC pellets is also in this direction, but with greater uncertainties because of the wider quality ranges of the material and the earlier step of technical development.

8.4.2 *Flexibility Through Biomethane*

The application of biomethane is another approach to a flexible bioenergy provision. This approach leads to a decoupling of the production and the use of the energy carrier, and thus allows various options for flexible bioenergy provision. This includes (i) short term, daily, weekly and seasonal flexible power provision (through long- to short-term storage and demand-based applications), (ii) providing a defined fuel for the transport sector and (iii) using the fuel in existing conversion units without technical adaptions. Hence, the main advantages of flexible energy provision from biomethane and bio-SNG can be compared with their similarity to natural gas.

Another flexibility option can be seen during production. Especially in combination with a power-to-gas-concept, see Chap. 2, where CO_2 from the upgrading process is needed to convert H_2 with solar or wind power to methane. Additionally, the product process itself can provide (with limits) more flexibility. Therefore, during the bio-SNG process, the type of gasifier and methanation unit combined with a change of load can influence the capacity. Fluidised bed and entrained flow gasifiers react quickly (in a matter of minutes) to a change in the load (0–100 %). Whereas fixed-bed gasifiers need a very long time to start and stop, a partial load can only be realised from 60 to 100 % [18, 19]. In any case it needs to be taken into account that a partial load has a much lower efficiency than a full load and in particular that plants in standby cool down rapidly without additional heating. For fluidised bed and entrained flow gasifiers it takes several minutes and in some cases hours to preheat the gasifier before the gasification reactions start. The main reason is the refractory material in the gasifier which has a high heat capacity and a limited temperature change velocity due to dilation and brittleness. The deployment time of methane synthesis is approximately 5 min, while the cold start time is in the range of hours [14]. The energy requirement for standby is about 1 % of max capacity. For implementation, more research on partial load operation of a synthesis plant is necessary.

During the biomethane process, a similar effect can be reached by a change of load. The production rate of the available default plant sizes (250 up to 2,800 m^3/h STP raw biogas upgrading capacity) can be modified. In most cases, the ability for down regulation is higher than for up regulation [2, 31]. Even the choice of substrates or operation mode of the digester can influence the productivity and thus the flexibility, see Chap. 6.

8.5 Conclusion

The **advanced solid biofuels** can contribute to a more flexible energy supply due to their favourable fuel parameters (e.g. low water content, easier to grind, high energy density or reduced volatile compounds) that not only enable an easier substitution of solid fossil fuels (with minor cost-incentive technical plant adaptations) but also a better process control of the biofuel in the conversion system and especially for thermally-treated fuels, much more flexible storage and transport options.

A successful instrument to improve the market implementation of thermally-treated solid biofuels can be seen by the success story of the worldwide production and trading system of wood pellets, which has clearly shown that there is a mutual interaction between the development of conversion technologies and fuels. The final potential of advanced solid biofuels can as such therefore not yet be described and is strongly dependent on the development of technologies.

Biomethane can contribute to a more flexible energy supply due to their similarity to natural gas as well as the flexible production and storage of the fuel. Therefore they can make a significant contribution to a flexible energy system, e.g. in the form of power-to-gas concepts or by meeting on demand energy provision in the heat, power or fuel sector. However, the demand for biogenic substitutions in these sectors is different.

The strengths and opportunities of these intermediates can be most greatly seen in logistics and usage, where there is easy and low price transportation through the already existing gas grid, new domestic and international markets and new applications such as shipping fuel [11].

Nevertheless, as is the case with all technologies, the production of intermediates not only shows strengths and opportunities but is also associated with weaknesses and threats. For example, the intermediate treatment processes are cost-intensive and often interconnected with additional demands for safety requirements in transportation and storage. The higher production costs have the potential of becoming economically feasible, by substantially reducing the cost and improving process efficiency. In addition, lower investment costs for transport, storage, conversion systems and maintenance services are to be expected as a result of the high-quality intermediate biofuels, if they can be implemented as a commodity biofuel on the market. Biogenic substitutes for natural gas are momentarily bound by transport to the gas grid, different quality standards between countries as well as high production efforts combined with high production costs. The technology is still developing however, which will ultimately lead to higher production efficiency. Furthermore, transport outside of the natural gas grid is possible under certain circumstances and will gain further importance in the future.

In the short-term, torrefied biomass will be available and biomethane is already commercially available through the biochemical process. However, the production of a biogenic substitute for natural gas from the thermochemical pathway, bio-SNG, is still under development and will be ready for the market in the long-term. It also appears that similar can be said for the implementation of the HTC process.

The resulting potential for a smart and future-based bioenergy system is versatile, promising and not yet even fully predictable. The historical development of the wood pellets market showed that especially supporting political and legal framework conditions can favour the way for a new biofuel implementation on national and international markets. Therefore, for the widespread market implementation and penetration of intermediate biofuels, they must be supported by an international fuel standardisation with a certification system along with the respective safety regulations.

References

1. T. Balling, M. Beil, A. Hauptmann, G. Reinhold, H. Seide, W. Urban, F. Valentin, B. Wirth, *Biomethaneinspeisung in der Landwirtschaft, KTBL-Schrift* (Kuratorium für Technik und Bauwesen in der Landwirtschaft e.V. (KTBL), Darmstadt, 2012)
2. F. Bauer, C. Hulteberg, T. Persson, D. Tamm, *Biogas Upgrading – Review of Commercial Technologies, SGC Rapport* (Swedisch Gas Technology Centre (SGC), Malmö, 2013)
3. R. Berger, Hein K R G Verfahrensübersicht: Synthesegaserzeugung aus Biomasse. FVS Fachtagung, 2003
4. J. Berner, Pelletswerk am Stück. Pellets Markt und Trend, Ausgabe 04/09, S. 26–28, Aug 2009
5. T.G. Bridgeman, J.M. Jones, I. Shield, P.T. Williams, Torrefaction of reed canary grass, wheat straw and willow to enhance solid fuel qualities and combustion properties. Fuel **87**, 844–56 (2008)
6. J. Daniel-Gromke, V. Denysenko, P. Sauter, K. Naumann, M. Scheftelowitz, A. Krautz, M. Beil, W. Beyrich, W. Peters et al., *Stromerzeugung aus Biomasse* (BMU, Leipzig/Berlin/Halle/Kassel, 2013)
7. DBFZ Internal analytic report 2013
8. M. Deutmeyer, D. Bradley, B. Hektor, J.R. Hess, L. Nikolaisen, J.S. Tumuluru, M. Wild, Possible effect of torrefaction on biomass trade. IEA bioenergy task 40, June 2012
9. DIN EN 14588–1, Solid biofuels – Terminology, definitions and descriptions; German version EN 14588:2010. Published 2011
10. EurObservÉR, Solid Biomass Barometer. Dec 2013
11. A. Florentinus, C. Hamelinck, A.V.D. Bos, R. Winkel, M. Cuijpers, *Potential of Biofuels for Shipping – Final Report* (ECOFYS Netherlands B.V, Utrecht, 2012)
12. FNR, *Leitfaden Biogasaufbereitung und -einspeisung* (Fachagentur Nachwachsende Rohstoffe e.V. (FNR), Gülzow, 2014)
13. GasNZV, Gasnetzzugangsverordnung (Gas Network Access Ordinance) – Verordnung über den Zugang zu Gasversorgungsnetzen, 2012
14. L. Grond, P. Schulze, J. Holstein, *Systems Analyses Power to Gas: A Technology Review*. Part of TKI project TKIG01038 (KEMA Nederland B.V., Groningen, 2013). http://www.dnv.com/binaries/dnv%20kema%20%282013%29%20-%20systems%20analyses%20power%20to%20gas%20-%20technology%20review_tcm4-567461.pdf
15. ISO 17225, ISO 17225- 1–8 Solid biofuels – Fuel specifications and classes. Part 8 under development, 2014
16. M. Kaltschmitt, Energie aus Biomasse, Grundlagen, Techniken und Verfahren. 2., neu bearb. und erw. Aufl (Springer, Berlin, 2009)
17. J. Kiel, R. Zwart, J. Witt, SECTOR – Production of solid sustainable energy carriers from biomass by means of torrefaction. International workshop on biomass torrefaction for energy, Albi (FR), 10. Mai 2012
18. M. Klemm, R. Wilhelm, A. Hiller, Untersuchung zur Einsatzmöglichkeit einer Biomassefestbettvergasung in einem virtuellen Kraftwerksverbund. 14. Symposium "Energie

aus Biomasse" Biogas, Flüssigkraftstoffe, Festbrennstoffe (Otti-Kolleg) Kloster Banz, 24–25 Nov 2005

19. M. Klemm, R. Wilhelm, A. Hiller, A Dezentrale Energieversorgung auf Basis einer Biomassefestbettvergasung in einem virtuellen Kraftwerksverbund. In: Tagungsband 37. Kraftwerkstechnisches Kolloquium. Dresden, 2005
20. M. Klemm, A. Clemens, R. Blümel, F. Kietzmann, Das HTC-Verfahren für Biomasse – ein Neubeginn für Kohle? 10. Jahrestagung Kraftwerke, Düsseldorf, 12 Nov 2013
21. H.A.M. Knoef (ed.), *Handbook Biomass Gasification* (BTG Biomass Technology Group BV, Enschede, 2005)
22. J. Kopyscinski, T.J. Schildhauer, S.M.A. Biollaz, Production of synthetic natural gas (SNG) from coal and dry biomass – a technology review from 1950 to 2009. Fuel **89**, 1763–1783 (2010)
23. F. Lettner, H. Timmerer, P. Haselbacher, Biomass gasification – State of the art description. Report for Intelligent Energy Europe, Grant agreement no. EIE/06/078/SI2.447511, 2007
24. I. Obernberger, A. Hammerschmid, *Dezentrale Biomasse-Kraft-Wärme-Kopplungstechnologien* (dbv – Verlag für die Technische Universität Graz, Graz, 1999)
25. I. Obernberger, G. Thek, *Herstellung und energetische Nutzung von Pellets: Produktionsprozess, Eigenschaften, Feuerungstechnik, Ökologie und Wirtschaftlichkeit*, 1. Aufl. (BIOS Bioenergiesysteme, Graz, 2009
26. A. Öhlinger, M. Förster, R. Kneer, Torrefaction of beechwood: a parametric study including heat of reaction and grindability. Fuel **104**, 607–613 (2013)
27. W. Rading, Vergasung als Option der thermochemischen Nutzung von Biomasse. TerraTec, Leipzig, 5–8 Mar 2007
28. C. Schön, H. Hartmann, Charakterisierung von Holzbriketts: Brennstofftechnische, physikalische und stoffliche Eigenschaften – eine Marktstichprobe. Berichte aus dem TFZ (Technologie- und Förderzentrum (TFZ), Straubing, 2011)
29. J. Shankar Tumuluru, S. Sokhansanj, J.R. Hess, C.T. Wright, R.D. Boardman, Review: a review on biomass torrefaction process and product properties for energy applications. Ind. Biotechnol. **7**, 384–401 (2011)
30. S. Steinbeiss, G. Gleixner, M. Antonietti, Effect of biochar amendment on soil carbon balance and soil microbial activity. Soil Biol. Biochem. **41**(6), 1301–1310 (2009)
31. Technische Universität Wien, Überblick über Biogasaufbereitungstechnologien zur Produktion von Biomethan, Bio-methane Regions (2012)
32. D. Thrän et al., Sustainable Strategies for Biomass Use in the European Context. Analysis in the charged debate on national guidelines and the competition between solid, liquid and gaseous biofuels. IE-Report 1/2006, Leipzig, 2006
33. J. Witt, Holzpelletbereitstellung für Kleinfeuerungsanlagen – Analyse und Bewertung von Einflussmöglichkeiten auf die Brennstoffqualität, Dissertation TUHH, DBFZ Report Nr. 14, Leipzig 2012
34. J. Witt, K. Schaubach, J. Kiel, M. Carbo, M. Wojcik, Torrefizierte Biomasse zum Einsatz im Kraftwerkssektor. 7th Rostocker Bioenergieforum, 20–21 Juni 2013

Chapter 9
The Potential of Flexible Power Generation from Biomass: A Case Study for a German Region

Philip Tafarte, Subhashree Das, Marcus Eichhorn, Martin Dotzauer, and Daniela Thrän

Abstract Energy scenarios and roadmaps indicate that intermittent renewable energy sources such as wind power and solar photovoltaic (PV) will be crucial to the power supply in the future. However, this increases the demand for flexible power generation, particularly under conditions of insufficient wind and/or solar irradiation. Among the renewable energy sources, bioenergy offers multiple end-use in the form of power, fuel or heat. Biomass-based power combines the advantages of being renewable, exceptionally CO_2 neutral and supporting demand-oriented production.

This chapter analyses four energy scenarios for Germany, focusing on the relevance of flexible bioenergy therein. Depending on how the scenarios are constructed, the range of biomass potential in the energy system is 1,180–1,700 PJ/a. The following sections of the chapter investigate the potential of flexible power generation from biomass on a regional scale (50 Hertz grid) starting with a description of the current state of bioenergy generation in the region and its potential for supplementary heat provision. We model the contribution of flexible biogas and solid biomass power using a minimization of daily residual load variance as a goal function. Two points in time are modeled – 2011 and 2030 to include the current and projected

P. Tafarte (✉) • S. Das • M. Eichhorn
Department of Bioenergy, Helmholtz Centre for Environmental Research – UFZ, Permoserstraße 15, 04318 Leipzig, Germany
e-mail: philip.tafarte@ufz.de; subhashree.das@ufz.de; marcus.eichhorn@ufz.de

M. Dotzauer
Deutsches Biomasseforschungszentrum GmbH – DBFZ, Torgauer Str. 116, 04347 Leipzig, Germany
e-mail: martin.dotzauer@dbfz.de

D. Thrän
Department of Bioenergy, Helmholtz Centre for Environmental Research – UFZ, Permoset Straße 15, 04318 Leipzig, Germany

Deutsches Biomasseforschungszentrum – DBFZ, Torgauer Straße 116, 04347 Leipzig, Germany

Bioenergy Systems, University of Leipzig, Grimmaische Straße 12, 04109 Leipzig, Germany
e-mail: daniela.thraen@ufz.de

D. Thrän (ed.), *Smart Bioenergy*, DOI 10.1007/978-3-319-16193-8_9

installed capacity from wind and solar PV. The results indicate that depending on the framework conditions, flexible bioenergy inclusion can reduce the daily variance in the residual load by >50 % compared to a non-flexible system. We conclude that flexible bioenergy has significant potential to contribute to balancing the power system with increasing shares of intermittent sources such as wind and solar PV.

9.1 Introduction

The previous chapters focused on the need for flexible bioenergy generation, resource availability, sustainability and environmental impact issues. This was extended by an overview of the available technologies and their potential for flexible energy generation from solid, liquid and gaseous biomass.

In this chapter, the potential for flexible power generation from biogas as well as solid biomass and its effect on the power supply system are demonstrated for a case study region – the area of the 50 Hertz transmission grid operator. The first section introduces some prominent examples of national energy scenarios. We focus on the role of bioenergy and the handling of fluctuations in the power supply within these roadmaps of energy transition. We demonstrate that there is still no silver bullet in sight at the moment and that several options remain possible. In Sect. 9.3 the study region with its current state of bioenergy use and its potential for supplementary heat use are illustrated. This forms the basis for the calculations in Sect. 9.4 which presents a numerical analysis of the contribution of biomass to flexible power generation in the study area followed by conclusions in Sect. 9.5.

9.2 Long-Term Potential for Flexible Bioenergy Generation

The biomass potential as discussed in previous chapters shows the upper limits for bioenergy provision. Further, it was explained that biomass is currently the only renewable source that contributes to all energy sectors e.g. power, heat and fuel and that bioenergy can be generated on demand with a short response time, enabling the balance of variable renewable sources (vRES) such as wind and solar photovoltaic (PV). However, from the scientific as well as the political perspective there is currently no consensus about the preferable end-use or function of biomass in the energy system.

Since the infrastructure of energy is fairly expensive and it is usually expected that it will serve for long time periods, e.g. up to 50 years for lignite or coal power plants, decision-makers usually base their decisions on sound scientific evidence. Scientific tools commonly used for the development and description of future energy systems are 'Energy Scenarios'. Energy scenarios at the national and/or international level have been developed and published since the 1970s [8]. By content, energy scenarios cover the impacts of individual political decisions on regional

Table 9.1 Overview of energy scenarios

Study title	Year	Name/Abbreviation	Institutes
Klimaschutz: Plan B 2050 – Energiekonzept für Deutschland [4]	2009	Greenpeace	Eutech Energie und Management GmbH
Modell Deutschland Klimaschutz bis 2050: Vom Ziel her denken [9]	2009	WWF	Institut für angewandte Ökologie ÖKO-Institut e.V., Prognos AG
Energieszenarien für ein Energiekonzept der Bundesregierung [12]	2010	BMWI	Prognos AG
			Energiewirtschaftliches Institut an der Universität zu Köln (EWI)
			Gesellschaft für Wirtschaftliche Strukturforschung mbH (GWS)
Langfristszenarien und Strategien für den Ausbau der erneuerbaren Energien in Deutschland bei Berücksichtigung der Entwicklung in Europa und global – Leitstudie 2011 [11]	**2012**	**Leitstudie**	Deutsches Zentrum für Luft- und Raumfahrt (DLR)
			Institut für Technische Thermodynamik, Abt. Systemanalyse und Technikbewertung
			Fraunhofer Institut für Windenergie und Energiesystemtechnik (IWES), Ingenieurbüro für neue Energien (IFNE)

and national energy systems up to changes and developments of the global energy supply system [8].

In order to get the full picture of the potential of bioenergy for flexible power generation, it is important to consider existing energy scenarios. Energy scenarios exist for Germany at the national scale [10, 14]. Some of them also consider a high share of fluctuating renewable resources; four of those recent and most prominent scenario studies (see Table 9.1) are briefly presented here.

9.2.1 Potential and Sector-Wise Distribution Under the Scenarios

Table 9.2 gives an overview of the expected sustainable primary energy potential of biomass under the scenarios. The results of the studies are relatively similar to one another in the range of 1,180–1,700 PJ/a, if import is excluded. This could be partially due to the fact that most of the scenarios (Leitstudie, Greenpeace and WWF) were basically based on the same fundamental literature [5].

The primary energy potential of bioenergy is distributed to different end-uses, separated into fuel for transportation, heat and the power supply. In 2010 about 30 % of the primary energy consumption was used for power, about 60 % for heat

Table 9.2 Sustainable bioenergy potential under the scenarios

	Leitstudie	BMWi	Greenpeace	WWF
Potential	[PJ/a]	[PJ/a]	[PJ/a]	[PJ/a]
Residue	800	NA	NA	700
Import	0	500	0	500
Others[a]	750	1,700	1,180	500
Total	**1,550**	**2,200**	**1,180**	**1,700**

NA not applicable
[a]E.g. energy crops, short rotation coppice, forest biomass

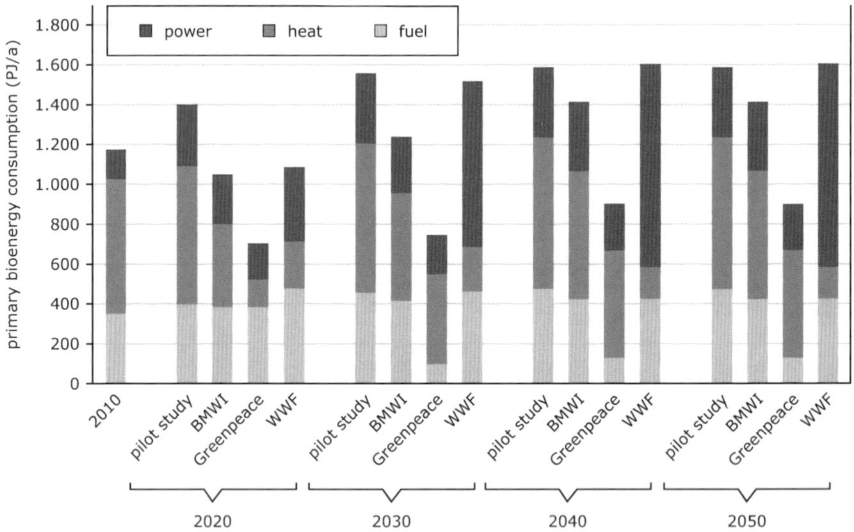

Fig. 9.1 Comparison of primary bioenergy consumption under relevant national scenarios (Based on personal communication with Julian Braun, DBFZ, 2013)

and 10 % for fuels [11]. However, under the scenarios, different development pathways with respect to the sectorial distributions of biomass are enumerated. This is basically due to a difference in the definitions of the sustainable application of biomass under framework conditions.

In Fig. 9.1, the contribution of primary bioenergy to the three sectors for a reference year 2010, as well as for the years 2020, 2030, 2040 and 2050 are displayed for comparison. Here, total and sectoral primary bioenergy consumption is compared under different scenarios. As it can be clearly seen in the figure, the scenarios differ with respect to power, heat and fuel consumption. The Greenpeace study which has a stronger focus on ecological aspects consistently allocates a lower (~ one-third) primary energy consumption of biomass compared to the other studies.

Against the above background, it can be concluded that only a small proportion of biomass is considered for power generation in the future. The following paragraphs clarify how the afore-mentioned studies deal with fluctuations and the specific role of bioenergy.

9.2.2 Flexible Power Generation Options Under the Scenarios

To compensate for fluctuations in feed-in from intermittent sources such as wind and photovoltaic, three options have been considered under the afore-mentioned scenarios: demand-side management, storage and instantaneous generation. Under the scenarios these options have been treated differently. In the following paragraphs, we discuss an instantaneous generation of power on demand, henceforth referred to as 'guaranteed capacity'.

Within the BMWI study, 50–70 GW guaranteed capacity has been calculated for the generation of balancing power. The largest contribution (~88–91 %) is provided by natural gas power plants and Carbon Capture and Storage (CCS) coal power plants. Biomass only contributes with 6 GW guaranteed capacity. However, full load hours of 6,500–6,800 h indicate that biomass plants operate in base load mode and are not managed for demand-oriented functioning.

As [11] shows, the expected guaranteed capacity is 68–77 GW. The main fraction of balancing power is foreseen to come from Combined Heat and Power (CHP) plants — both fossil-fuel driven as well as those fired by gaseous biofuels such as Biogas or Biomethane. Two pathways are considered in [11] with respect to the use of biogenic gaseous energy carriers. Firstly, the feed-in into the existing natural gas net for power and heat generation in large CHP plants and secondly the on-the-spot conversion to power whenever balancing power is required. For the latter option, modifications of existing bioenergy plants are necessary e.g. an increase in the installed capacity and storage capacity. The effects of a flexible on-the-spot conversion concept on the power system will be highlighted in the case study in Sect. 9.4.

The Greenpeace study mentions the challenges of tackling fluctuations in wind and solar PV, but it does not provide explicit quantifications. The WWF study calculates a guaranteed capacity of 59–61 GW depending on the scenario assumptions. This guaranteed capacity is separated into contributions from renewable sources plus imports (23–27 GW), conventional sources (mainly natural gas) and storage (34–36 GW). It does not explain however the exact contributions of the individual renewable energy sources.

Conclusively, a comparison of the studies on various scenarios shows that the role of biomass is more diverse than that of the other renewables but has not been discussed in detail along with its implications. The role of biomass in these studies is seen as ranging from base load operation mode for mainly heat and power production to a flexible source for balancing fluctuations in intermittent renewable sources (e.g. wind and solar). To use the specific advantages of bioenergy for balancing power grids, more information about the effect of flexible generation from biomass is needed. For such a smart bioenergy provision to be integrated into the overall energy system it is important to consider the regional framework condition, including the current state of bioenergy plants in operation, the demand for power and heat and the electricity grid situation. In the following (Sect. 9.3), we present a discussion of the current state of bioenergy plant distribution and the heat potential thereof followed by Sect. 9.4 which gives an example of the system effects of flexible power generation from biomass as a case study of the 50 Hertz Grid operator area in eastern Germany.

9.3 Regional Aspects of Bioenergy

This section introduces the study region for which flexible power generation from bioenergy has been modelled in the following sections. The study was conducted in eastern Germany. Geographically, the region covers seven German federal states (Hamburg, Berlin, Brandenburg, Saxony, Saxony-Anhalt, Thuringia, Mecklenburg-Western Pomerania) covering a total area of 109,340 km^2. The area is operated by 50 Hertz Transmission GmbH, which functions as the Transmission System Operator (TSO) serving about 21 % of the German population [15] (Fig.9.2).

In a classical energy supply chain, centralized systems played a major role. However, a high level of integration makes centralized systems vulnerable to changes within the supply chain. Decentralized systems, as a model of supply infrastructure, are less vulnerable to the availability of remote generation and transmission networks [6]. Furthermore, the demand for flexible power generation in a changing energy system with a high proportion of intermittent renewable sources (wind and solar PV) reaches the limits of possibilities offered by centralized fossil fuel power plants. Centralized systems are usually developed to operate at nominal capacity throughout the year which does not allow them to follow the high load gradients demanded by the feed-in of intermittent renewable sources. Flexible bioenergy is therefore emerging as a good option due to two main advantages (i) utility in decentralization mode and (ii) the ability to follow load gradients (e.g. power generation from biogas). However, the introduction of flexibility concepts to the bioenergy sector is also highly dependent upon regional or local aspects of energy production. Spatial aspects of current infrastructure are also crucial for establishing flexible energy systems at regional scales.

In the selected 50 Hertz region, the total number of plants (including biogas, solid biomass and biofuel plants) is 1,773 (2011). The total installed capacity in the region is ~1,365 MW with an average of 769 kW. The spatial distribution of these plants is shown in Fig. 9.3 while Table 9.3 shows the distribution of plants.

CHP plants primarily serve electricity production, however, heat, which is a by-product of the process may also be used e.g. for district heating. When introducing

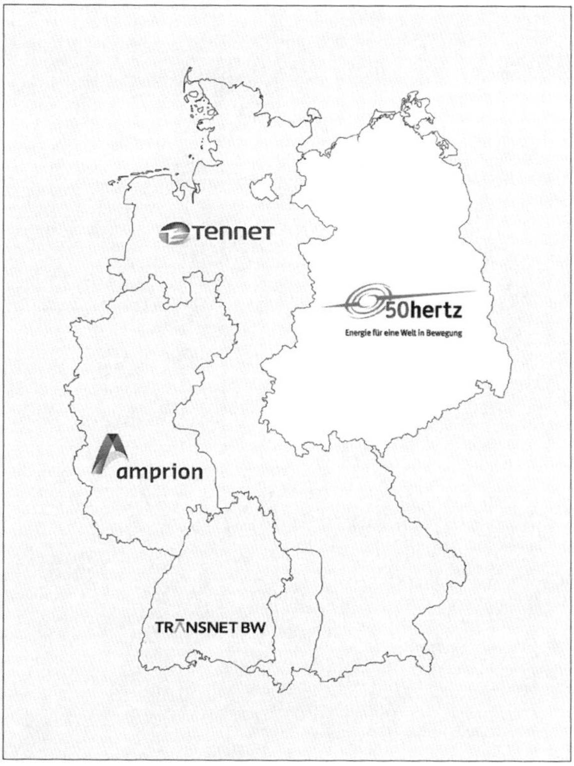

Fig. 9.2 Transmission Network Operators in Germany

flexible options it is relevant to address the potential of district heating from biogas, since both flexibility and heat demands have temporal dimensions. Further, the spatio-temporal consideration of heat sinks in the design and implementation of flexible plants may be valuable in reducing storage requirements.

A further investigation into the biogas facilities installed in Saxony showed that currently these plants are driven by electricity demand and provide base load, thereby using only a minor proportion of the produced heat [7]. Results indicated that the total heat supply potential from biogas plants in the region is around 290 GWh (i.e. ~15 % of the heat demand in the region could be potentially fulfilled from bioenergy plants). The study identified a strong limitation due to a lack of demand centers around the plants with respect to housing infrastructure. About 40 % (194 GWh) of the heat that was theoretically available for supply faced geographical constraints for further use in district heating systems, because the plants are located too far away from the demand centers. However, in certain cases heat provision can act as a constraining factor for flexible power generation.

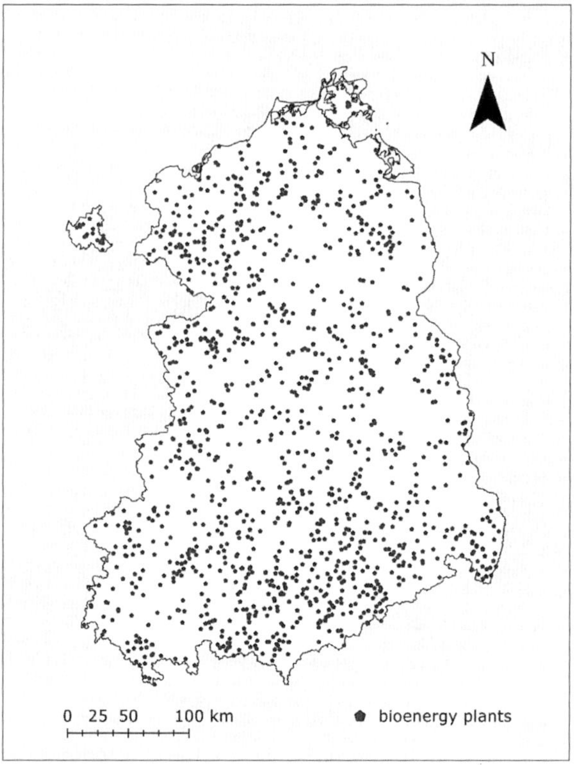

Fig. 9.3 Regional distribution of bioenergy plants

Table 9.3 Distribution of bioenergy plants in the 50Hertz grid region

Range of installed capacity (kW)	Number of plants	Total installed capacity (MW)
<500	1,006	307
501–1,000	643	391
1,001–3,500	81	137
3,501–45,000	10	46
5,001–10,000	17	108
>10,000	16	373
Total	**1,773**	**1,364**

Based on [1]

9.4 Complementing Variable Renewable Energies with Flexible Bioenergy

In the following paragraphs, the effect of flexible power generation from bioenergy to balance fluctuations in the electricity supply is demonstrated as a case study. To assess the balancing potential on fluctuations from variable renewable energy sources such as wind and solar photovoltaic (PV) as well as fluctuations of power demand, flexible bioenergy power generation is modelled for one of the four German power transmission grids, operated by 50HertzTransmission GmbH (50Hertz). Based on 3-year time series data for demand and feed-in from wind and solar, the effect of flexible bioenergy power production can be compared to current non-flexible bioenergy power generation. Residual Load (RL), calculated as the difference between the demand and supply from wind and photovoltaic forms the basis for modelling bioenergy power provision.

Since both demand and feed-in from wind and solar PV fluctuate, the compensation of the RL has to balance out these fluctuations for a stable power supply system. In contrast to non-flexible power production from bioenergy, flexible bioenergy generation is expected to contribute to the balancing of the power system, especially in cases of substantial shares of fluctuating renewable energy sources without any major power storages, e.g. large pumped hydro-storage systems.

Apart from assessing the effects of either flexible or non-flexible bioenergy power generation we also provide a scenario for the projected increase in installed capacities from wind and solar PV for 2030. The framing conditions for 2030 (installed capacities, annual energy power production and power demand) are adapted versions of [11]. Table 9.4 presents a comparison between 2011 and 2030 parameters. Two bioenergy technologies (biogas and solid biomass) are modelled, because they account for more than 90 % of the installed bioenergy capacity in the 50Hertz grid (see Sect. 9.3).

9.4.1 Model Description

Based on the time series data from 2009 to 2011 [3] the RL is calculated from the capacity given in Table 9.4 by a proportional scaling of the feed-in from wind and solar PV power plants. Feed-in from all bioenergy plants was simulated for two modes: (i) non-flexible power production and (ii) flexible power production. The

Table 9.4 Scenario conditions for the case study

	Year 2011		Year 2030	
	Capacity (CAP)	Annual energy production	Capacity (CAP)	Annual energy production
	[MW]	[TWh/a]	[MW]	[TWh/a]
Wind	11,719	18	17,979	41
Solar	4,070	3	10,005	9
Bioenergy	1,460[a]	9	2,435	15
Solid biomass	861[a]	5	1,552	9
Biogas	599[a]	4	883	6
Total	**17,249**	**30** (~36 % of demand[b])	**30,419**	**65** (~76 % of demand[b])

[a]Based on the average demand from 2009 to 2011 of 84 TWh, capacity for 2011 from 50Hertz plant data [2], capacity for 2030 derived from [11]

[b]Demand for 2030 falls by 10 % as projected by [11], 6.8 TWh of energy from wind and solar are considered to be excess energy in 2030

differences between non-flexible versus flexible power generation from bioenergy have been studied with a minimum temporal resolution of 1 h. The results from these simulations were compared to estimate the contribution of either mode to the reduction in fluctuations of RL.

To capture the effect of non-flexible bioenergy power production on RL, a constant feed-in of bioenergy is subtracted from the original RL resulting in a new RL after compensating for bioenergy ($RLB_{(t)non\ flex}$). In this case the value of "const" is equal to 1 so that no flexible operation of the bioenergy power generation capacity is possible.

$$RLB_{(t)nonflex} = RL_{(t)} - const * \left(CAP_{solid} + CAP_{biogas}\right) \tag{9.1}$$

CAP = installed capacity of either solid or biogas plants.

In the case of flexible power generation, the power production is enabled to adapt to RL fluctuations by allowing the optimization algorithm to modulate the power generation. This is realized by introducing the modulation factor "m" which scales the power generation of the capacity from bioenergy plants, so that a minimization of daily variances in RL is achieved [13]. This modulation forces power generation from bioenergy to contribute to the balancing of the power supply and demand by shifting flexible power generation from times of lower RL to times of higher RL.

On a daily basis, power from bioenergy is provided at times of high RL and reduced at times of relatively low RL throughout the time series from 2009 to 2011. As the flexible operation is modelled in sequence for the two different technologies (solid biomass and biogas), the resulting RL after the introduction of flexible bioenergy generation from $RLB_{flex\ solid}$ and $RLB_{flex\ biogas}$ is $RLB_{flex\ combined}$:

$$RLB_{(t)flexsolid} = RL_{(t)} - m_{(t)solid} * CAP_{solid} \tag{9.2}$$

$$RLB_{(t)flexcombined} = RLB_{(t)flexsolid} - m_{(t)biogas} * CAP_{biogas} \tag{9.3}$$

$$min\,variances\left(m_{(t)solid}; m_{(t)biogas}\right) = \sum_{t=1}^{24} RLB_{(t)flexcombined} \tag{9.4}$$

The "variances" as a function of the two modulation factors "m(t) solid" and "m(t) biogas" are subject to minimization in this modelling for the 24 h of each day throughout the time series.

The details of the parameterization of the model are described in the following paragraphs. The key technical parameters are provided in Table 9.5.

The operation of solid biomass and biogas capacity is modelled in sequence to improve the combined effect of the different flexibility potential from both bioenergy technologies. Setting the more dynamic biogas capacities second after the less dynamic solid biomass capacities should ensure that the characteristics of both technologies are not operated in a conflicting way but rather in a complementary

Table 9.5 Technical parameters for the flexible operation of power generation from solid bioenergy and biogas plants

	Bioenergy technologies	
	Solid biomass	Biogas
Modulation of power output	0.5–1 in 2 h time steps (0.5–1.2 for 2030)	0–2 in 1 h time steps
Operational constraints	Constant daily energy production	Constant daily energy production
	No storage limitations for input materials affecting operation	On-site biogas storage equivalent to 12–24 h in biogas production
	Reduced daily production (−20 %) during summer from April to October	Reduced biogas production (−25 %) on weekends assuming feeding management
Energy production	Annual Energy Production (AEP) remains constant for either non-flexible or flexible operation. AEP from biomass in 2030 taken from [11]	

interplay. The parameterization and operation of either technology is explained in the following:

1. Solid Biomass Plants: The combined installed capacity from solid biomass plants is first modulated from 0.5 to 1 (0.5 to 1.2 in the 2030 case) for every 2 h time step of the time series, meaning that the combined installed capacity from solid biomass plants is multiplied by the modulation factor and subtracted from the RL time series. A modulation factor of 0.5 is applied as the minimum modulation factor as heat demand from CHP production and standard conversion technology currently does not allow for a power output below 0.5 or 50 % of the rated power. The lower heat demand in summer is taken into account by a reduction in daily energy production by 20 % compared to the operation during winter.
2. Biogas Plants: The combined installed capacity from biogas plants is modulated from 0 to 2 on the basis of the RL after the feed-in from solid biomass plants (as above). The maximum modulation factor of 2 points out that the installed capacity can provide twice the power output to allow for a more flexible production compared to the current almost constant modulation factor of 1. The constraint of a maintained overall daily production together with the modulation factors of 0 to 2 implies a maximum storage capacity on site for 12–24 h, although no detailed storage modelling is performed.

   On weekends with a generally lower power demand, the daily production of biogas and consequently power and heat production is reduced by 25 % assuming a feeding management of the biogas digester.

Since the most common operation mode in bioenergy plants is CHP, the given parameterization of the modeling allows for bioenergy plants to operate throughout the year to maintain a high utilization of heat without the necessity to deploy increased heat storage facilities.

9.4.2 Results

The results presented in this section correspond to the capacity provided in Table 9.4 (1,460 MW in 2011 and 2,435 MW in 2030 for bioenergy). The calculations were based on the time-series of 2009–2011 for RL and feed-in from wind and solar PV. The combined results from a flexible operation of solid bioenergy and biogas capacity are presented in Table 9.6.

The results demonstrate that flexible bioenergy production improves maximum and minimum RLs and variance in daily RL for both 2011 and 2030 cases. The flexible bioenergy generation enables a significant reduction of the variance in daily RL by 56 % for 2011 and 54 % for 2030 compared to the non-flexible reference. This leads to a significant reduction in load variations for the remaining non-renewable power generation system. It reduces the maximum RLs compared to a non-flexible operation by 7 % (2011) and 12 % (2030) compared to the 2011 level for non-flexible operation selected as the reference (100 %). As a result, this directly contributes to reductions in power plant capacity to provide the remaining residual power production. Likewise, the minimum RL or excess power is reduced, avoiding power production at times when power generated from wind and solar already completely meet the demand for power.

A closer look at the temporal operation patterns for the flexible bioenergy plants reveals that the modulation of power output adapts to the short-term production patterns of variable renewable energy sources as well as fluctuations in demand. As shown in Fig. 9.4, the power production of the solar PV installations

Table 9.6 Overview of the results from simulated flexible and non-flexible bioenergy power generation in the case study

	Year 2011		Year 2030	
	Non-flexible operation	Flexible operation	Non-flexible operation	Flexible operation
Variance in daily residual load	100 %	Reduced by 56 %[a]	100 %[a/**]	Reduced by 54 %[a]
Maximum residual load (deficit power)	12,499 MW (100 %)	11,651 MW (reduced by 7 %[a])	10,343 MW (100 %[a])	9,047 MW (reduced by 12 %[a])
Minimum residual load (excess power)	3,980 MW (100 %)	3,352 MW (reduced by 16 %[a])	13,536 MW (100 %[a])	12,538 MW (reduced by 7 %[a])
Bioenergy power production in times of excess power	176 GWh/a	118 GWh/a	11,010 GWh/a	10,021 GWh/a
Avoided bioenergy power production in times of excess power by flexible operation	–	58 GWh/a (reduced by 33 %)	–	990 GWh/a (reduced by 9 %)

[a]Percentages compared to "non-flexible" values

[**]The high levels for 2030 figures are caused by fluctuation from increased vRES capacities

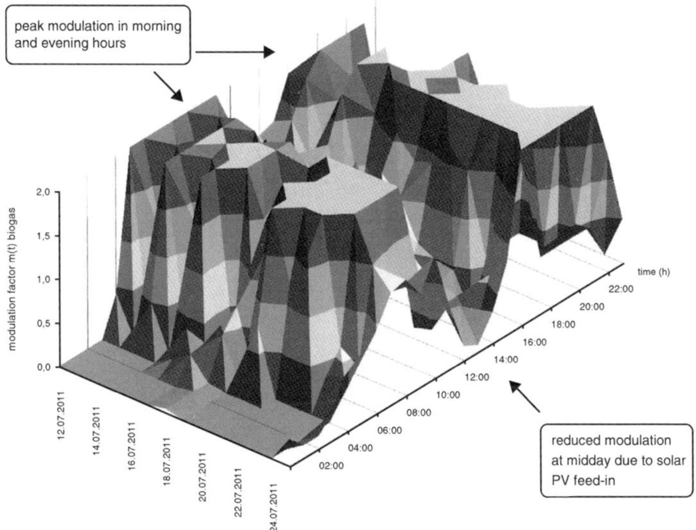

Fig. 9.4 Example for the modulation of biogas power generation in high insolation conditions

(4,070 MW/10,005 MW in 2011/2030) is responsible for the reduced RL at midday in high insolation conditions, leading to a low utilization of flexible bioenergy power production. Bioenergy power generation is instead shifted to provide maximum power production in morning hours and late evening hours when high demand cannot be met by solar PV.

Figure 9.5 depicts seasonal patterns of the effect of flexible bioenergy production on average daily RL before (solid lines) and after (dotted lines) the feed-in from flexible power production. The resulting RL shows a significant reduction in the average daily RL amplitude compared to the original RL.

Figures 9.6 and 9.7 show the duration curves for the simulated time series projected for 2011 and 2030. These duration curves are created by ordering all hourly RL values of the 3-year time series in a descending order, so that the highest RL value is located on the very left of the graph and the lowest value on the right side.

As shown by the duration curves, the flexible operation of bioenergy plants in the modelled set-up allows for a limited shift of power production (grey area between solid lines of the RL) from times of lower RL on the right side of the duration curve to times of higher RL on the left. This not only helps to reduce negative RL (excess power) from renewable energy, but also reduces maximum positive RL (deficit power), enabling a reduction in non-fluctuating plant capacity, which is currently mostly driven by fossil fuels.

The comparison of the duration curves of the RL in 2011 and 2030 reflects how a substantial increase in capacity for wind and solar power has an impact on the RL distribution. The overall duration curve shifts so that instead of a mere 120 h per year of negative RL (excess power) for 2011, over 2,000 h of negative RL per year are calculated for 2030. The maximum negative RL (excess power) over the 3 year

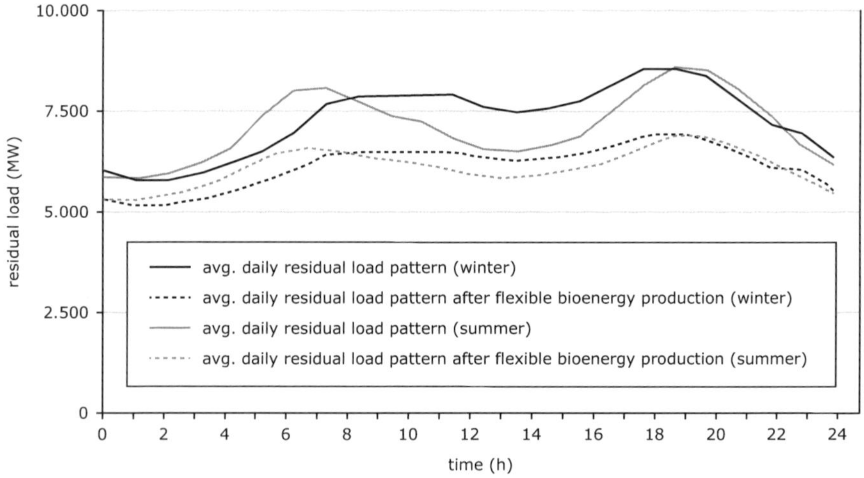

Fig. 9.5 Reduced amplitude in average daily RL after flexible bioenergy power generation, differentiated for winter and summer (2011 installed capacity) (Note: typical reduction of RL at noon in summer time, caused by high solar PV feed-in)

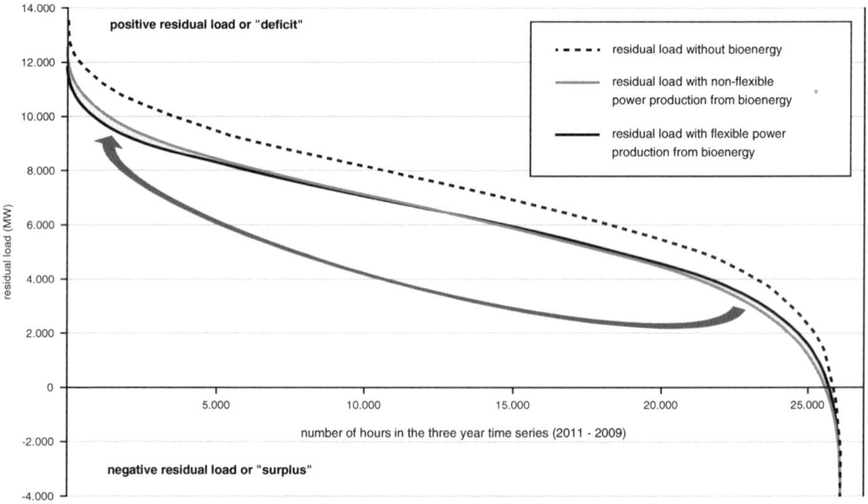

Fig. 9.6 RL curves for the 50Hertz grid network with flexible and non-flexible bioenergy (2011 installed capacity, 3-year reference period)

time-series increases from 3,980 MW (2011 capacities) to 13,536 MW (2030 capacities) (see also Table 9.6). This reflects an overall increase in capacity of variable power production from wind and solar PV. For flexible bioenergy, the consequence is that the demand for flexibility to complement these increased fluctuations will likewise increase. For example, power production from biomass has to be increasingly shifted over longer periods when prolonged periods of high power production from wind and solar are already serving the power demand.

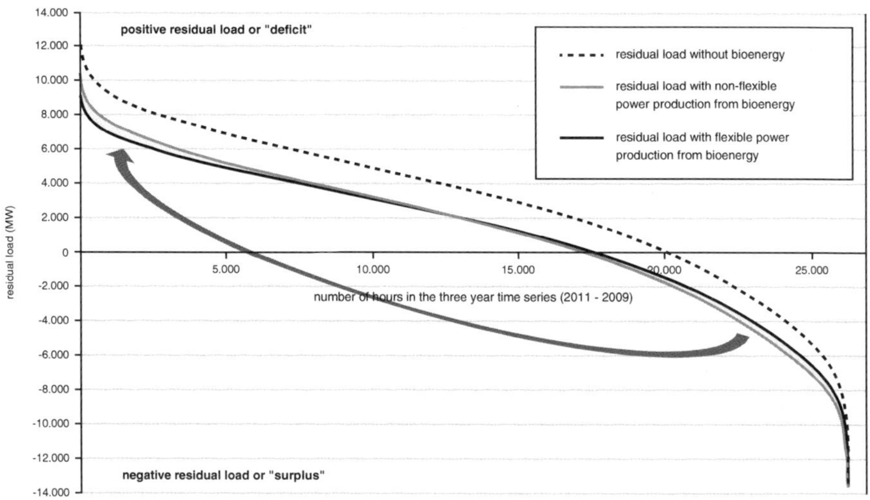

Fig. 9.7 RL curves for the 50Hertz grid network with flexible and non-flexible bioenergy (2030 installed capacity, 3-year reference period)

Of the 15,000 GWh/a of energy from biomass in 2030, about 3,500 GWh/a are shifted from times of low RL on the right side of the graph to times of high RL. Of these 3,500 GWh, about 990 GWh/a are shifted from times of negative RL so that bioenergy is not produced in times of fulfilled demand by wind and PV but shifted instead to times of positive RL. The remaining 2,510 GWh/a are produced even though wind and PV provide sufficient power to supply demand.

9.4.3 Discussion

This chapter investigated the potential of flexible bioenergy as an option for balancing fluctuations in the power grid resulting from load patterns and increasing vRES shares. The results from this regional case study indicate that flexible bioenergy can contribute positively towards balancing power grids.

Based on available renewable energy scenarios, an increase of vRES capacity (wind and PV) from 2011 to 2030 was modelled for the Eastern German region. The limited installed capacity of bioenergy in this case study (1,520 MW/2,435 MW from bioenergy in 2011/2030) is far too low to fully balance fluctuations of vRES capacity (15,789 MW/27,984 MW of Wind and solar PV in 2011/2030). However, the introduction and operation of flexible bioenergy capacity to balance fluctuations in RL (as shown in this case study) through the hourly modulation of capacity to minimize daily RL variance has been verified as an effective measure to balance short-term fluctuations. The simulation revealed a reduction in variability of more than 50 % compared to the reference case of non-flexible operation for both 2011

(56 %) and 2030 (54 %) (see Table 9.6). Modest improvements from flexible operation were identified in terms of maximum excess power and deficit power over the course of the 3-year simulation period, providing additional benefits for the power grid.

According to the simulations presented here, in 2011 the proportion of excess power or negative RL in the system was negligible (176 GWh/a). The modelling results indicate that 58 GWh/a of bioenergy generation could be shifted to compensate positive RL. By the year 2030 an increased share of vRES (see Table 9.4) and excess energy (11,010 GWh/a) in the system is expected. As for the modelling results, from the 3,500 GWh/a that would have been generated from biomass without a flexible operation in times of excess, 990 GWh/a could be shifted by flexible operation. To unlock the remaining 2,510 GWh/a and enable an additional shifting of bioenergy in 2030, greater flexibility is needed.

Therefore, these results indicate that flexible bioenergy provision in the short-term is an effective measure to balance a renewable system (with negligible excess energy), but that future (e.g. 2030) flexibility options will need to be complemented by additional flexibility options and further investments, i.e. in gas and heat storage.

Both, solid biomass power plants and biogas plants were taken into consideration, but with different assumptions about their flexibility. Solid biomass power plants are constrained in their modulation range (0.5–1.2). Although this limits their flexibility potential, power production may run at nominal capacity for long time periods as long as a sufficient stockpile of biomass is available for any addition to the base modulation factor of 1. By contrast, biogas plants with increased generator capacity can be modulated more dynamically than solid biomass plants (modulation factor 0–2). One of the factors that currently restricts flexible generation is the limited capacity to freely regulate biogas production as it is based on anaerobic digestion processes (see Chap. 5).

In general, flexible biogas plants with biogas storage on-site of 12–24 h are well suited to complement the daily production pattern of solar PV at times of high solar irradiance. As no such regular, semi-deterministic production pattern exists for wind power which has a greater dependence on high and low pressure weather systems over Germany prevailing typically for more than 12–24 h, the selected modelling setup is not sufficient to address the means of balancing long-term fluctuations from wind energy. One option to address this shortcoming is to link biogas plants to the natural gas grid to make use of the huge storage potential of the existing gas grid (see Chap. 5). This can overcome the limitations of on-site storage for biogas to cope with the long-term variability in RL.

While some inflexibility is presumably caused by restrictions of the modelling in this case study, as the applied optimization routine is restricted to daily load fluctuations and falls short of inter-daily shifting of power production from bioenergy. However, the flexibility of the biomass technologies which are used in the modelling as well as the operational constraints from combined heat and power operation limits the flexibility in the setup that was investigated.

It is worth mentioning here that this study used RL as a 'known input parameter (from the data)' which by contrast is only partially predictable in real-time plant operation. However, the above results for 2011 and 2030 are based on a set of ex-post data (measured/reported/calculated) specific to the 50Hertz region, implying that the optimization results and conclusions hold true for the set of input data used. The main benefit of using this approach is that it clearly illustrates the advantages of 'flex' bioenergy over using non-flexible bioenergy. Furthermore, results from the 3-year time-slice (RL and RES feed-in) and the applied modelling in this study provides a range of the calculated potential of bioenergy flexibility, allowing for a reduction in daily RL variance of up 56 %.

This case study strongly indicates that the adoption of flexible bioenergy has the potential of supporting the energy transition in Germany. In addition to demonstrating the technical options for flexible bioenergy as presented here, a detailed techno-economic feasibility assessment should be carried out to get the full picture. Innovations and/or adaptations to technologies need to be integrated into the current modelling process as and when required. Flexible bioenergy also needs to be adequately supported by policy, especially by specific incentives that promote flexible bioenergy and framed by sustainability requirements for the feedstock supply. In summary, flexible bioenergy does not necessitate additional bioenergy production but focusses on improving the use of bioenergy that has already been produced, while quantifying the future role of bioenergy in the energy sector can greatly benefit flexible bioenergy provision.

9.5 Conclusion

A transformation of bioenergy provision from a stand-alone provision to integrated systems can be realized on a regional level. A deeper analysis of the East German region showed that it is possible to start changing the existing installation to support the transition of the energy system in the immediate future. By enabling a flexible power provision from biomass, this will result in a higher value of the electricity provided, a reduction in the overall RL to be covered by fossil fuels, while neither the demand for biomass nor the combined heat supply are significantly altered.

For a description of future pathways towards a renewable energy supply, the options for flexible power provision from biomass should be included. So far, the available scenarios do not or not fully consider these and therefore assume higher RLs as well as more energy from fossil fuels. There is a need to adapt these scenarios –not only in terms of modified bioenergy provision but also in terms of economic effects: flexible bioenergy provision calls for much greater technical effort and leads to higher specific provision costs while the reduction of RL has a clear potential for cost reduction in the mid-term.

From the calculations in the case study, an increased negative RL can be expected while at the same time increasing the potential of bioenergy to reduce the fossil RL. Hence, in the long term, a flexible power generation from biomass has the potential of becoming a major contributor to the power supply. However, the results also show that the capacity of power provision from bioenergy is far too low to fully balance fluctuations of the vRES capacity. Consequently, if renewable power provision is to be directly integrated into the energy system, the optimization of power provision from bioenergy is only one aspect. Hence, this case study can be regarded as a starting point for a systematic optimization, which will inevitably lead to some additional potential and challenges for future developments:

1. Today the contribution of flexible power provision from solid biofuels is limited due to the currently installed technologies. Whereas new technologies will be available that support future flexibility –especially the provision of synthetic natural gas (SNG) and/or the power generation in gasification units –with the potential of a wider modulation. In this case, the flexibility of solid biofuels and biogas might be comparable in 2030. This has not been considered in the case study, because so far it cannot be estimated when and how those technologies will be in place on the market.
2. The case study focused on short term flexibility with a shift of electricity provision within 24 h (modulation rate of 0.5–2). Increasing this modulation and also including longer term flexibility might provide additional potential to balance fluctuations in the power system. The previous chapter showed how additional technical options are being developed to provide mid- and long-term flexible power.
3. Not only the electricity generation from biomass needs to be optimized with a view to system integration, but also the fluctuating energy carrier wind and solar PV can contribute to reduce fluctuations in RL, by taking into account spatio-temporal feed-in patterns and advancements in wind and solar PV technology [15]. Hence, the additional installation of renewable power capacity should be framed by integrated planning, considering those aspects as soon as possible.
4. Heat provision also has some additional effects on flexible power provision: on the one hand, CHP concepts require dedicated heat supply concepts for mid- and long-term flexible power provision. On the other hand, the availability of excess energy might lead to additional power-to-heat concepts as a second pillar for heat supply in an energy system mainly based on renewables. Both aspects have not been tackled here and need further investigation.

In terms of an efficient reduction of greenhouse gases, today's possible "no-regret-options" to reduce fossil-based power generation by adapting the existing biogas plants should be realized soon. Therefore, adjusted framework conditions are necessary to make investments in the additional power conversion unit (second CHP-engine) of the biogas plant feasible. This will be discussed in detail in Chap. 10.

References

1. 50HertzTransmissionGmbH, Jahresabrechnung 2011, http://www.50hertz.com/de/166.htm. Retrieved 28 Mar 2014
2. 50HertzTransmissionGmbH, EEG-Anlagenstammdaten, http://www.50hertz.com/de/165.htm. Retrieved 20 Nov 2013
3. 50HertzTransmissionGmbH, Zeitlicher Verlauf der EEG-Stromeinspeisung, http://www.50hertz.com/de/file/EEG-Bewegungsdaten_2011.zip. Retrieved 20 Nov 2013
4. K. Barzantny, S. Achner, S. Vomberg, *Klimaschutz: PlanB 2050 – Energiekonzept für Deutschland (Langfassung)* (Greenpeace e.V. EUtech Energy and Management GmbH, 2009), p. 154
5. BMU. Bundesumweltministerium [Hrsg.]: J. Nitsch, W. Krewitt, M. Nast, M. Pehnt, G. Reinhardt, M. Fischedick et al., *Ökologisch optimierter Ausbau der Nutzung erneuerbarer Energien in Deutschland.* (Untersuchung im Auftrag des Bundesministeriums für Umwelt/Naturschutz und Reaktorsicherheit (BMU), DLR Stuttgart, IFEU Heidelberg, WI Wuppertal, 2004
6. F. Bouffard, D.S. Kirschen, Centralised and distributed electricity systems. Energy Policy **36**, 4504–4508 (2008)
7. S. Das, M. Eichhorn, M. Hopfgarten et al., *Spatial analysis of the potential of district heating from existing bioenergy installations in Germany.* 20th European biomass conference and exhibition, Milan, 2012
8. C. Dieckhoff, W. Fichtner, A. Grunwald, A. Voß, M. Schönfelder, P. Jochem et al., *Energieszenarien. Konstruktion, Bewertung und Wirkung – "Anbieter" und "Nachfrager" im Dialog* (Karlsruher Institut für Technologie (KIT), Karlsruhe, 2011), p. 164
9. A. Kirchner, M.E. Schlesinger, B. Weinmann et al., *Modell Deutschland Klimaschutz bis 2050: Vom Ziel her denken* (WWF Worldwide Fond for Natur, Prognos AG, Öko-Institut e.V. Institute for Applied Ecology, Basel/Berlin, 2009), p. 533
10. T. Kronenberg, D. Martinsen, T. Pesch et al., *Energieszenarien für Deutschland: Stand der Literatur und methodische Auswertung* (Forschungszentrum Jühlich, Institut für Energie- und Klimaforschung, Systemforschung und Technologische Entwicklung (IEK-STE), Jühlich, 2011)
11. J. Nitsch, T. Pregger, T. Naegler et al., *Langfristszenarien und Strategien für den Ausbau der erneuerbaren Energien in Deutschland bei Berücksichtigung der Entwicklung in Europa und global* (Deutsches Zentrum für luft und Raumfahrt (DLR)/Fraunhofer Institut für Windenergie und Energiesystemtechnik (IWES)/Ingenieurbüro für neue Energien (IFNE), Stuttgart/Kassel/Teltow, 2012), p. 273
12. M.E. Schlesinger, P. Hofer, A. Kemmler et al., *Energieszenarien für ein Energiekonzept der Bundesregierung.* Projekt Nr 12/10 (Prognos AG, Energiewirtschaftliches Institut an der Universität zu Köln (EWI), Gesellschaft für wirtschaftliche Strukturforschung mbH (GWS), Basel/Köln/Osnabrück, 2010
13. M. Schreiber, Bewertungskriterien einer optimierten Energieversorgung regionaler Verbünde mit hohen Anteilen erneuerbarer Erzeugung. Zeitschrift für Energiewirtschaft **36** (4), 257–265 (2012)
14. N. Szarka, F. Scholwin, M. Trommler, F. Jacobi, M. Eichhorn, A. Ortwein, D. Thrän, A novel role for bioenergy: A flexible, demand-oriented power supply. Energy **61**, 18–26 (2013)
15. P. Tafarte, S. Das, M. Eichhorn, D. Thrän, Small adaptations, big impacts: Options for an optimized mix of variable renewable energy sources. Energy **72**, 80–92 (2014)

Chapter 10
Conclusion and Outlook

Daniela Thrän

Abstract In a nutshell, smart bioenergy can be described as the process of optimizing individual technologies and plants to an optimized contribution of bioenergy technologies to the overall energy system and infrastructure with the benefit of providing additional services from bioenergy. The focus of this book is on the conceptual approaches and the technical potential for developing different biomass provision routes towards more flexibility. This requires conversion plants with units that can be controlled with precision and well adapted to short reaction times, with a partial load function of the conversion process and additional storage facilities.

Power provision from biomass is one application, where increasing flexibility can be expected in Germany over the next 5 years when electricity from wind and photovoltaic will become more important. Due to the specific frame conditions of power provision, the demand for flexibility in this sector is expected to be very challenging, requiring reaction times of only a few minutes to provide positive or even negative energy to balance grid stability. Beside the specific German case, flexible power to increase the grid stability can be necessary due to different reasons and is required in many countries of the world. Highly flexible heat provision in small scale combustion units is not so much an issue at the moment, but is expected prospectively to be due to an increasing supply of heat from solar systems and/or heat from excess energy from wind and photovoltaic (power to heat). Fuels for transportation are also expected to change in the years to come. Furthermore, the increased availability of fluctuating wind and solar power will provide excess energy during certain periods. Basically speaking, the excess electrical energy can be converted into thermal or chemical energy and meet some of the demand for heat or fuel consumption. As a result, some of the flexibility needs can be shifted between the different sectors. To enable technologies to fulfil the additional demands for smaller and more flexible bioenergy provision, the availability of advanced intermediates is

D. Thrän (✉)
Department of Bioenergy, Helmholtz Centre for Environmental Research – UFZ, Permoset Straße 15, 04318 Leipzig, Germany

Deutsches Biomasseforschungszentrum – DBFZ, Torgauer Straße 116, 04347 Leipzig, Germany

Bioenergy Systems, University of Leipzig, Grimmaische Straße 12, 04109 Leipzig, Germany
e-mail: daniela.thraen@ufz.de

D. Thrän (ed.), *Smart Bioenergy*, DOI 10.1007/978-3-319-16193-8_10

a core issue. This includes further development and market implementation of advanced solid biofuels as well as biomethane.

Challenges on the road to becoming more flexible do not only occur from the technical options and limitations but also from the elements of the supply chain, including sustainable feedstock provision, the implementation of flexible conversion concepts and the demand from the renewable energy market. With regard to the holistic system approach, three pillars for smart bioenergy systems can be identified: (i) an additional demand for smaller application units in terms of energy provision from biomass, (ii) the necessity to have improved technologies providing the desired products in small units and (iii) and new concepts of system integration – including the energy system but the coupled production of materials energy carriers from biomass as well.

It is only through the combined actions of different stakeholders that flexible bioenergy can be implemented successfully. A stepwise approach to achieving flexibilisation has to be designed and a careful consideration of the directed transition of the related energy systems is imperative. The bigger picture of such an upcoming energy supply system is the combined provision of heat, power and fuels based on different renewable energy carriers. Moreover, smart bioenergy needs to be coupled with future bio-economy approaches, providing materials and energy from the limited feedstock. The book does not go into detail here but many of the elaborated technical and managing elements, such as the sustainable feedstock base, designed intermediates and controlled conversion processes in production networks are necessary for flexible bioenergy provision and for advanced bio-based material production within a future bio-economy.

10.1 Main Insights and Lessons Learnt from This Book

10.1.1 Smart Bioenergy in a Nutshell

The transition of the energy system towards renewable resources is a key target of national and international agreements. The first major steps in this direction were made in the last decade. In the long term a greatly reduced role of fossil fuels is expected. According to this transition, bioenergy needs to play a more integrated role in the energy system – providing controllable energy that can fill the gaps of fluctuating renewables (wind and sun), balance seasonal shifts and provide liquid and gaseous biofuels for specific applications.

In a nutshell, smart bioenergy can be described as the process of optimizing individual technologies and plants to an optimized contribution of bioenergy technologies to the overall energy system and infrastructure with the benefit of providing additional services from bioenergy. The demand for these new concepts is driven by the changing markets for power, heat and transport fuels, which have very different frame conditions in different countries. Even if most of today's energy markets are still not demanding flexible bioenergy, the expected demands for bioenergy and the potential adaptions of today's bioenergy provision systems are being discussed. Furthermore, the potential effects for the overall energy system are often demonstrated using Germany as an example, because it is currently experiencing a dynamic transition of the energy system towards a renewable resource base, in particular for its power supply system in the near future, but also ultimately planned for its heat provision.

10.1.2 Improving Technologies and Concepts for a More Flexible Bioenergy Provision

This book mainly focuses on the conceptual approaches and the technical potential for developing different paths to biomass provision towards greater flexibility. This requires conversion plants with units that can be controlled with precision and well adapted to short reaction times, with a partial load function of the conversion process and additional storage facilities.

Power provision from biomass is one application, where flexibility can be expected in Germany over the next 5 years. Due to the specific frame conditions of power provision based on wind and photovoltaics, the demand for flexibility in this sector is expected to be very challenging, requiring reaction times of only a few minutes to provide positive or even negative energy to balance grid stability. Additionally, climate conditions, such as sun and wind, frequently change the demand of the remaining residual load to be supplied by additional renewable sources. With this specific demand for flexible renewable power Germany might have a specific frontrunner position, but there is also a more general requirement for flexible power to back up the local and regional electricity grid stability due their electricity infrastructure. Flexible power provision from biomass has to provide both very short-term controllable units as well as additional storage and conversion facilities for times with higher residual load provision. There are various power generation concepts from biomass available with different kinds of conversion principles, different levels of technical readiness, start-stop-behaviour and ramping (load change) ability.

- For the thermo-chemical conversion of solid biofuels, current flexibility is realised by reducing full load operation of the plants with limited load change rates. In the future advanced concepts can be expected that will provide even greater flexibility, relying on gasification as a thermo-chemical conversion process due to the flexible handling of the gaseous intermediate energy carriers. While the basic units of these future systems are already available, their complexity requires some further research work. The preliminary larger plants of this kind are expected to be installed by 2025 under favorable conditions.
- With the provision of power from biogas and biomethane, flexibility is currently derived from decentralised combined heat and power units with short reaction times. Due to the option of gas storage, power provision can be systematically shifted during the day to cover inconsistent residual loads that remain after the feed in of fluctuating electricity from wind and sun. Reliable and remote-controlled units are able to provide controlled power, while also compensating

current for grid stability. The crucial components that define the flexibility of the process are (i) control systems for gas utilization, (ii) adequate gas storage capacity and (iii) a sufficient process control system for the biological process. Strategies for increasing the flexibility of a given system aim to feed the digestion process with easily degradable substrate, i.e. sugar beet or the run-off from silage storage systems, to achieve shorter reaction times of the gas production process. For new plants, fermentation units come into question, which allow a high degree of flexibility and controllability in biogas production. The technical components that enable a flexible operation are already available. However, full-scale experience is still lacking.

- Power from liquid biofuels also offers interesting options for flexible operation, but generally speaking only a minor provision of power from liquid biofuels is expected due to the limited resource base for liquid biofuels and their outstanding properties in the transport sector.

Flexible power provision in combined heat and power units can be limited by the heat provision concepts of the plants. The relevance of this limitation is expected to be different for the many concepts of biomass-CHP applied today in Germany and could potentially be solved by additional heat storage systems in many cases. For all concepts however, flexible power provision is expected to require additional installation and control units and to reduce the power provided. This requires transparent market conditions and operation modes and business concepts adopted for each individual plant.

Heat provision in small-scale combustion units provides by far the highest share of bioenergy. Looking at German conditions, flexible heat provision from biomass is not so much an issue of today but can be expected prospectively from an increasing supply of heat from solar systems and/or heat from excess energy from wind and photovoltaic (power to heat). Furthermore, the specific heat demand in the residential heating sector will decrease dramatically so that future supply systems will be faced with the concept of flexibilisation and smaller heat provision units. For flexibilisation again reliable and remote-controlled systems with short reaction times will be necessary. A wide range of technologies are available and the overall challenge is to combine the conversion system, the control units, the heat buffer and alternative renewable supply systems in the most efficient way. One precondition for a more flexible heat supply from biomass is the availability of well-controlled processes with automatic feeding systems and defined solid biofuels. With regard to the expected reduction of the conversion unit size, advanced fuels such as hydrothermocal converted biomass or torrefied wood pellets together with very light and highly adjustable combustion systems will ensure a smooth operation. Due to a changing ratio of power and heat demand in housing towards power, the use of micro-CHP-units will probably become an alternative as opposed to heat generators on their own.

The fuels used for transport are also expected to change in the years to come. In contrast to heat and power however, the renewable alternatives to gasoline, diesel and kerosene these days are exclusively based on biomass. The use of wind and solar energy in the mobility sector requires a substantial transition in driving systems and infrastructure and significant technology development for accumulators, alternative storage systems as well as engine design. The potential for the provision of greater quantities of so called power-to-liquid or gaseous fuels from renewable power is also an option that will be relevant in 15–25 years. The integration of transport fuels from biomass into the energy system is therefore surrounded by the questions: (i) how will green house gas emissions reductions in the transport sector be managed in general and (ii) what kinds of biofuels will play which roles in which transport sectors (e.g. road, rail, water and aviation) in the medium and long term. These are questions that are not only open in the case of Germany, but for many other nations around the world as well. Many different concepts for bio-based transport fuels have been discussed. Generally speaking, the flexibility of all of the concepts concentrates on raw materials and other input factors such as auxiliaries (reaction media, catalysts) and on plant operation in terms of the main and by-product ratio within a given product portfolio. The most important flexibility of biofuels stems from their excellent storage and transport properties, meaning that a very flexible application for end energy provision is possible. From today's perspective we expect the same quality of flexible provision of available biofuels (biodiesel, bioethanol) and future biofuels (such as BtL and other thermo-chemically generated biofuels). Future flexible polyproduct or polygeneration plants will attempt to produce the most profitable products according to market fluctuations. Those so called biorefineries can take advantage from a tremendous smaller scaling compared to fossil refineries and a better regional integration. They will have to meet the major challenges in designing polyproduct concepts while taking into account the optimal trade-off between operational flexibility and capital cost, which might also lead to a stronger consideration of material provision from biomass with liquid biofuels as by-products.

10.1.3 From Plant Optimization Towards the Optimization of the Overall Energy System

Technologies that provide flexible power and heat from biomass are a starting point for the overall optimization of the energy system. Flexible provision consists of adopted conversion plants in combination with storage systems for biomass and different biofuels. Biomass storage is well established with the option of providing solid, liquid and gaseous biofuels irrespective of the harvesting and collection time of the biomass. Biofuel storage is also used to provide the end energy irrespective of biofuel conversion. For liquid biofuels this is state of the art, but for gaseous biofuels further concepts need to be established, including on-site storage at the conversion plant (i.e. extra biogas storage systems at biogas CHP plants) and the use of the natural gas grid as storage. Moreover, adopted heat storage systems can also be necessary in some cases. Flexible bioenergy provision requires a systematic optimization of the conversion and storage concept as a whole.

Furthermore, the increased availability of fluctuating wind and solar power will provide excess energy during certain periods. For the case study in Eastern Germany this excess energy will increase dramatically under current scenarios for the renewable energy supply until 2030. Generally speaking, the excess electrical energy can be converted into thermal or chemical energy and meet some of the demand for heat or fuel consumption. So-called "power-to heat" uses the excess energy for heat provision in smaller and larger applications (i.e. houses, accommodation, industry, district heating etc.); the concepts are already available today.

So-called "power to gas" concepts aim to convert the excess energy into hydrogen or other renewable gases. Here first demonstration plants are currently being built. "Power-to-liquid" concepts finally include the further processing of the renewable gas to different fuels; this is still in the conceptual stage. In the future, for the German energy market it is expected, that the sectoral analysis of power, heat, transport and gas markets will only be able to deliver half of the picture because all market segments are expected to merge. As a result, some of the needs for flexibility can be shifted between the different sectors. From today's perspective this can be regarded as a second step in the transition. This will also affect the future bioenergy provision concepts in many ways, which cannot be holistically conceived today. Two examples of potential integration are as follows:

Combining power-to-gas and biogas: Biomass, especially biogas, can be integrated into power-to-gas concepts by providing the renewable carbon source for the provision of renewable gases or liquids as chemical energy storages. Hydrogen from excess energy conversion can be used to increase the methane output of the biogas process. In this case, the hydrogen is converted to methane by adding carbon dioxide – a major byproduct within the biogas production process. This methanisation can take place as a thermochemical process by using catalysts or as a biological process. For a biological-based conversion different options for process design are currently being discussed (i.e. injection of hydrogen into the biogas process through membranes, direct injection of the gas, separate fermentation tower for gas conversion).

Combining power-to-heat and biomass heat: additionally, with a more renewable energy supply, the opportunity will arise to generate heat at times when the electricity rate is low from surplus renewable electricity. This can be realized using existing technologies (heating elements). The realization of such concepts will strongly depend on the specific frame conditions, i.e. the availability of excess energy and the effort for the additional components.

For bio-based transport fuels, system integration does not so much focus on the integration into the energy system, but much more on the combined provision of materials and biofuels in biorefinery approaches, where the starting point of plant concepts and optimization still needs further development.

Finally, energy efficiency from the end user's perspective, changes in consumption patterns and the adaptation of wind and solar towards wider ranges of energy provision are all necessary. These options have not been analyzed in the previous chapter but preliminary calculations show that the effects of a flexible bioenergy provision can be increased dramatically by the systematic optimization of all renewable power provision systems (see Ref. [12]).

10.1.4 Advanced Bioenergy Carriers for Efficient Flexible Provision

To enable the technologies to fulfil the additional demands for smaller and more flexible bioenergy provision, the availability of advanced intermediates is a central issue. Here on the one hand, solid biofuels are relevant for the further development of smaller conversion units that are easier to control in the heat and power sector, but also for a more efficient conversion of biomass to transport fuels. The worldwide production and trading system of wood pellets has clearly shown that there is a mutual interaction between the development of conversion technologies and fuels, ultimately leading to new application technologies and application fields in smaller and better controlled conversion units (pellet stoves and pellet boilers). The ultimate potential of advanced solid biofuels thus still cannot be described. Bioenergy carriers with clearly defined, standardised properties still have to be developed step by step. In addition to wood pellets, the provision of thermo-chemically treated pellets such as torrefied pellets, or hydrothermal converted biomass are promising intermediates to support the provision of high quality solid biofuels from different sustainable available feedstock (Fig. 10.1).

The conversion of biomass to biomethane provides the option of substituting natural gas in many different applications for energetic and material purposes. Biomethane in particular is an interesting fuel for flexible, small-scale application with high emission standards such as combined heat and power supply in residential and office buildings in urban areas. For the transport sector, biomethane is another option compared to different liquid biofuels. Here again however, clearly defined, standardised properties can fulfil specific requirements for certain transport options (i.e. aviation). Hence, synthetic biofuels can be regarded as a more promising option for a renewable-based energy supply.

10.1.5 The Importance of a Sustainable Feedstock Base

It is important to understand that smart bioenergy is not only based on the adjustment of the compounds of the conversion plants but also on the further development of the biomass provision. Furthermore, the transition of the energy system strongly depends on the sustainable provision of biomass, taking into account greenhouse gas emissions, emissions causing acidification and nitrification but also very local effects such as soil quality (including e.g. soil organic carbon), water use, biodiversity as well as social aspects.

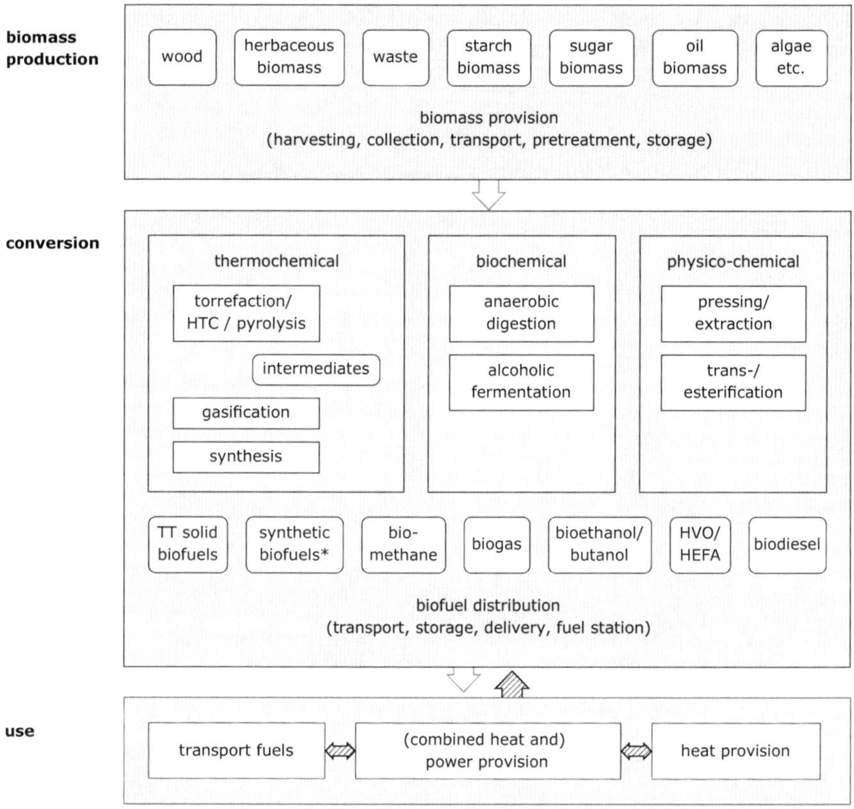

Fig. 10.1 Advanced bioenergy carrier as an interface between biomass feedstock conversion and an integrated renewable energy supply (*TT* thermochemically treated; * Fischer-Tropsch Diesel, DME, Biohydrogen etc.). The *grey arrows* show new relationships in future energy systems

Under the current debate, the additional availability of biomass for energy provision is uncertain and driven by strong and dynamic factors such as the development of crop yields and livestock numbers, population growth, per capita food consumption, land use conservation and change as well as climate change. The global potential of residues, by-products and waste is around 5–10 GJ per capita and year. This leads to a minimum biomass potential in the range of 50 EJ/a worldwide, which is more or less the amount of biomass that is used for energy provision today. The minimum additional bioenergy potential can be gained from improved biomass conversion of so-called traditional biomass, which includes at least a doubling of today's energy provision from biomass. This minimum potential mainly consists of woody and herbaceous biomass and different waste influxes.

Biomass potential from energy crops are assessed very differently showing results in the range of 50–200 EJ/a worldwide [1, 13]. Faced with the competition from food and fiber, driven by population growth, nutrition behavior and crop yield development, the future bioenergy potential is also influenced by the assessment of impacts from land use change from energy crop plantations and their associated greenhouse gas emissions, by the overall discussion on boundaries, i.e. in terms of biodiversity loss [10], and other environmental effects. For example, it can be expected that biomass as a renewable resource will be assessed more differentiated, also in IPCC calculations [5]. Changing strategies for biomass use for energy crops can therefore follow. Hence, it is not only a question of availability of arable land for cultivating energy crops, but also of the expected impacts from their use.

Some of those effects are considered in the ongoing sustainability debate. Existing schemes for sustainability certification differ significantly with regard to the indicators they consider and thus the completeness of their standards. Additionally, they are only binding for certain bioenergy carriers. Since major aspects for environmental sustainability such as the protection of natural areas are of high relevance for agricultural and forestry production in general, a more coherent sustainability framework should include all kinds of biomass (food, feed, fibre, fuel) and should also be realised in international agreements. This is a precondition to provide sustainable feedstock not only from a regional level, but from international markets as well. Higher demands for biomass are expected for the feedstock supply of biorefineries, because the flexibilisation of those concepts is limited. From today's perspective it is difficult to imagine which kind of biorefinery concepts are the most promising. Tradable sustainable feedstock is a key element, also for smart bioenergy.

Sustainable feedstock also leads to additional requirements for conversion technologies from biomass to bioenergy carriers, to make sure that the wider range of feedstock properties can be transferred to standardized fuel qualities. The access to a wider range of feedstocks might improve feedstock availability in a certain region and stabilize the feedstock prices, but it will not change the overall biomass potential for energy use dramatically.

10.1.6 Future Demands

Even if the need for flexible bioenergy in the long term seems obvious, its implementation has not yet been fully prepared. Barriers and opportunities not only arise from the technical options and limitations, but also from aspects of the supply chain, including feedstock provision, the implementation of adopted conversion technologies and the demand from the renewables market. With regard to the holistic system approach, flexible bioenergy requires both an implementation of the concepts and integration into the future energy system. This leads to different factors that are necessary for the future demand, as shown in Table 10.1.

Sustainable Feedstock Flexible bioenergy might reduce the specific biomass demand for the single conversion unit in many cases. This will increase the requirements for fuel quality and logistics. Nevertheless, the demands of the renewable energy system are not in line with the regional biomass potential. Even without an

Table 10.1 Future demands along the bioenergy supply chain for flexible bioenergy provision with regard to implementation and integration

Factors of the supply chain	Implementation of flexible bioenergy concepts	Integration of flexible bioenergy concepts
Sustainable feedstock	Availability of adapted biofuels and logistics	Availability of coherent sustainability standards
Adapted conversion technologies	Successful demonstration of technologies and concepts	Access to market and support for implementation
Renewable energy market	Assessment of (additional) effects of bioenergy flexibilisation	Assessment of the future demand for flexible bioenergy

overall growth of biomass use, feedstock supply gaps can be expected and need to be filled by internationally-traded biomass or intermediates. Hence, major challenges in terms of social, ecological, technical, logistical and economic aspects of international bioenergy trade will have to be overcome.

Conversion Technologies Technologies and concepts for smart bioenergy are at different levels of technical readiness and need further research and development. Successful demonstration is a precondition for market access. Besides the elementary technical aspects, there are also economic constraints. One big drawback of a flexible mode of operation is the increase in production costs, due to the fact that for the same amount of energy, a major sum of money is needed for upgrading investments. In Germany, the first incentive was created in 2012, providing a marked premium for bioenergy and an additional premium for flexible power supply from biogas [2]. The initiated shift of more than 1,000 biogas plants to the flexible operation until the end of 2013 provided numerous experiences of how to operate the units and how to control flexible bioenergy provision from pooled plants [11]. Due to increased efforts for flexibilisation and without any clear advantages, as the prices for CO_2 certificates, gas and oil, the market does not provide sufficient refinancing for system integration. From the integration perspective, future provision concepts for flexible bioenergy require full market access. This can include the permission to provide controlled power for grid stability to capacity reserve and other system services which have to be provided for a secure electricity supply. It can also include standards for biomethane injection into the natural gas grid. A flexible provision of heat and transport fuels is realized by the storage of the biofuels. There is no system-oriented implementation on the market yet. Specifically for bio-based transport fuels in the future, integration into biorefinery concepts will become more important than the integration into the energy supply system only. In the long term, market access for flexible bioenergy needs to be settled widely, including specific regional frame conditions but also – where necessary – on an international level, i.e. by international agreements and standards for solid, gaseous and liquid biofuels.

The Renewable Energy Market The intended effect of flexible bioenergy is the support of the overall transition of the energy system towards renewables within a wider vision of a sustainable resource use. Therefore an assessment of the flexibilisation with regard to the energy system and also to the overall environment is absolutely necessary during the market implementation phase. Preliminary insights in the case of Germany have shown that the effect of flexible power provision from bioenergy plants can reduce the residual load from fossil fuels to a certain degree with the technical adaptation of today's available technologies, but with higher shares of renewables, additional factors will be necessary. Even here however, the bigger picture is required: In many cases flexibility will mean reducing the full load production of a conversion unit from a basic load to a partial load or seasonal operation or changing the technical concepts to smaller conversion units to be operated in modules. We did not investigate the cost of smart bioenergy in detail, but we expect that in comparison with the modern bioenergy production the provision costs a single energy unit will increase in many cases. Ideally, those costs are covered by payments for the additional services, which can be manifold. The clarification of assessment approaches for system services in energy and product systems based on renewables is in a very early stage and will remain as one of the open points at the end of this book. To find out the added value from a macro-economic perspective, and to adopt the concepts that will optimize the added value are major questions for system integration.

Last but not least there are different promising starting points for flexibilisation: Due to the dynamic transformation of the electricity sector, additional qualities will be required in regions with a high share of fluctuating wind and solar energy provision in the short term. A regional integration of bioenergy, and especially combined heat and power will require appropriate framework conditions for energy system planning. By comparison, changes in building infrastructure and their heating systems are rather slow in many countries around the world, with i.e. exchange rates of 1–2 % for buildings and 2–3 % for heating systems in Germany. Therefore, establishing a higher flexibility of biomass heat generators and renewable heating systems with biomass will be a process that will last several decades. Finally, for the provision of biofuels for transport, the overall strategy needs to be discussed before adopting certain concepts.

10.2 The Way Forward: Actions Required

Only the combined actions of different stakeholders will enable a successful implementation of flexible bioenergy. A stepwise approach has to be designed that considers the directed transition of the related energy systems. Below, we discuss in more detail (based on the German case study), what will be needed from each stakeholder group to make this happen.

10.2.1 Policymakers

The transition of the energy supply towards renewables is one of the biggest challenges of the twenty-first century. National stop-and-go policies have proven to disrupt the market significantly. To frame the transition by stable conditions for all stakeholders along the provision chain is one of the most difficult governance problems. These framework conditions include:

1. One of the major driving forces for the energy transition is climate protection. There is no doubt about the need of coherent, long-term working policy instruments to implement the polluter-pays-principle for greenhouse gas emissions. Policy has to find ways of making current approaches (emission certificates) work or support them with additional measures (i.e. CO_2 taxes). Additionally, transparent and consistent monitoring systems should be established on an international level to minimize the risk of confusion and facilitate the discussion between policy and society.
2. For a successful transition of the electricity sector, full access for renewables to the electricity market and the different kinds of services to compensate the currently controlled power is necessary. National policy framework for electricity markets should provide this access for renewables in general. Power supply systems with a high share of renewables might require new instruments to compensate for capacities. Furthermore, an adopted design of market introduction instruments for renewables (feed-in tariffs, quotas, biddings) in terms of technical requirements (possibility of remote control, start-stop-behavior) and incentives to provide flexible renewable energy is necessary. Flexible bioenergy provides additional services for integrated energy supply systems with large shares of renewables. It is a well-known fact that the landscape demand for renewables (wind, solar, biomass) is substantially increased compared to fossil fuels [7]. Both aspects require a stronger consideration of regional contexts for energy systems planning and the optimization of bioenergy within those systems. Due to the shifts between the final energy markets, the consideration of the regional context has to include heat, power and transport fuel supply and the connecting infrastructure (electricity grids, gas grids etc.). A successful implementation requires fast working, efficient planning instruments and legislation framework, which are not available yet in most countries.
3. The technical and conceptual development of smart bioenergy needs to be holistically encouraged in national and international research strategies. This includes applied research activities on (i) components for process control, (ii) improvement of intermediates and storage, (iii) full scale experience on operation under varying loads for all technologies and market readiness levels.
4. Liquid biofuels are state of the art today and are highly flexible in storage, distribution and end use application. Nevertheless, in the transport fuel sector, the role of bioenergy is limited. Under current conditions, the use of biofuels is not necessarily linked to specific needs from the energy system (i.e. difficulties to find alternative renewable supply options) or other value added effects along the supply chain (i.e. providing transport fuels with excellent fuel qualities or increasing access to transport fuels in certain regions of the world). For liquid biofuels over the next decade, additional flexibility for plant operation is not so much an issue, but more so the strategic production and application of the fuels. Additionally, there is a strong need to clarify the long term strategies for biomass utilization

for the transport sector on a national and international level. This is especially necessary for the integrated production of those biofuels in future multiproduct biorefineries in a bio-based economy.

10.2.2 Academia

Although this book has provided numerous insights into the potential future role of bioenergy, many questions are still open. Scientific methods and approaches are needed for technological development and demonstration of the different components for smart bioenergy, but also for the requirements and options of the overall renewable energy system into which they have to be integrated:

1. To picture potential future options, more specified coherent assessments of the environmental, economic and social aspects are needed. Clearfication is also necessary for the time frame of the transition: at what stage of the transition does the change towards smart bioenergy make the most sense, and how to ensure that the shift towards bioenergy is possible at this time and not locked by frame conditions that have already been implemented.
2. To close the gaps in the sustainability assessment of bioenergy provision chains, some scientific answers have not yet been found. For example, there are different scientific views on the greenhouse gas effects of indirect land use change and the carbon accounting from forest biomass [6]. Besides the generation of methods and results, the discussion process of the scientific community needs to be improved to provide robust assessments even under uncertainties.
3. To provide flexible bioenergy as an additional service for future energy supply systems, the potential positive effects have been described in this book and the expected additional costs have been discussed. Today, only rough calculations for the overall cost aspects of smart bioenergy are available – this is true for the technical concepts but especially for the macroeconomic effects of smart bioenergy aspects in an energy supply based on renewable resources.
4. To deliver scientific tools and instruments for necessary technological developments by further developed transient process modeling (e.g. flowsheeting or computational fluid dynamics) as well as control and measurement technologies.
5. To figure out new business cases for highly flexible bioenergy and biomass use considering the special aspects of centralized and decentralized concepts and cross-over effects between the energy sectors. This also includes future flexible polyproduct or polygeneration plants that will try to produce the most profitable material and energy products altering production according to market fluctuations.
6. Last but not least from a German perspective the smart bioenergy approach is one element of a knowledge-based demand-oriented provision of bio-based products and services for sustainable economic systems [14]. In the German

debate it is strongly linked to research from social science, dialogues with society and monitoring systems [4]. Exemplarily, the further development of bioenergy can be taken as a research case to investigate the challenges and opportunities of such approaches.

10.2.3 Non-governmental Organizations (NGOs)

The flexible bioenergy debate has to be embedded in larger systems. This provides opportunities and risks for stakeholders and for the environment. Both require participation and the development of safety fences with regard to many different aspects, i.e. sustainable feedstock, location of conversion plants and infrastructure, connecting smart bioenergy into energy-saving strategies. A key issue is the implementation of sustainability indicators for biomass on an international level. NGOs can support the stepwise expansion of such indicators also for non-energy biomass utilization (see also Sect. *10.2.4 Business, action field 2*).

Besides these well-known topics for NGOs, flexible bioenergy on a regional level can present new business models such as "citizen energy plants" or "virtual power plants". With regard to the overall challenges and opportunities of the transition (see Sect. *10.2.2 Academia, action-field 5*), participatory implementation activities might also profit from active debate and concrete proposals from NGOs in this new playing field.

10.2.4 Business

On the one hand, the bioenergy industry needs clear framework conditions to invest in and manage new bioenergy approaches. On the other hand, awareness of feedstock availability and competition is a precondition for successful business models [6]. For the implementation of flexible bioenergy, industry activity should focus on quality standards for the feedstock as well as intermediates:

1. International sustainability standards for biomass are implemented but need to be further developed stepwise, in terms of the dimensions included (environmental, economic and social), and in terms of the sectors included (fuel, food, feed, fibre). Certain initiatives for those standards can be stated at the national and international level (i.e. Forest Steward Council (FSC), Programme for the Endorsement of Forest Certification Schemes (PEFC), Round table on Sustainable Palmoil (RSPO), International Sustainability and Carbon Certification (ISCC)), but so far there are only utilisation paths for specific materials that are certified. The effort to establish sustainability standards and certification systems for dedicated bioenergy opened a door to land-use related production. NGOs should seize this opportunity to widen the discussion to other

materials – and to learn from bioenergy for sourcing feedstock for other bio-based production systems within a sustainable framework.
2. The integration of flexible bioenergy needs specification-driven energy carriers for the national and international trade of bioenergy carriers. Therefore, the international technical standards for solid biofuels and natural gas substitutes (biomethane) have to be expanded in terms of new application fields.

Novel opportunities for new businesses can be expected in the field of process control, the communication between processes and plants, the management of virtual plants and supply chain management. In Germany in particular, there is the additional challenge of upgrading existing CHP plants from biogas and solid biofuels to a more flexible electricity provision within the existing control and support schemes.

10.3 Closing Remarks: A Vision of the Future Renewable Energy Supply: Smart Bioenergy and a Bio-based Economy

Renewable energy is increasing tremendously in all sectors all over the world [9]. The combination of different renewable resources and a stronger role of the decentralized, integrated energy supply are key factors for expected energy system changes [3]. Furthermore, by pooling different renewable power generation plants (wind, solar and biogas), the services for stabilizing the electricity grid can already be provided today [8]. Even if large scale units still run until the middle of the century in many countries, it is unlikely that they will be reinstalled and their manifold system services will soon be replaced by renewables.

The bigger picture of the upcoming energy supply system is a combined provision of heat, power and fuels based on different renewable energy carriers. Bioenergy can provide various units for a regionally adapted renewable energy provision using different renewable power provision systems, controlling and storage options as well as grid connections for renewable gases (Fig. 10.2). This can include:

- controlled power for power grid stabilisation
- the capacity to cover the electrical residual load as well as heat demand
- upgrading and storage of excess electrical energy through (semi)biogenious gases
- renewable heat provision support in larger settlements, public buildings, industry etc.
- heat provision in regions with a very low heat demand i.e. individual buildings or rural settlements
- biofuel provision for specific applications (i.e. heavy duty vehicles, ships, rail and aviation or as a synergy with electromobility)
- technology approaches developed for flexible biorefineries producing materials and biofuels in coupled processes

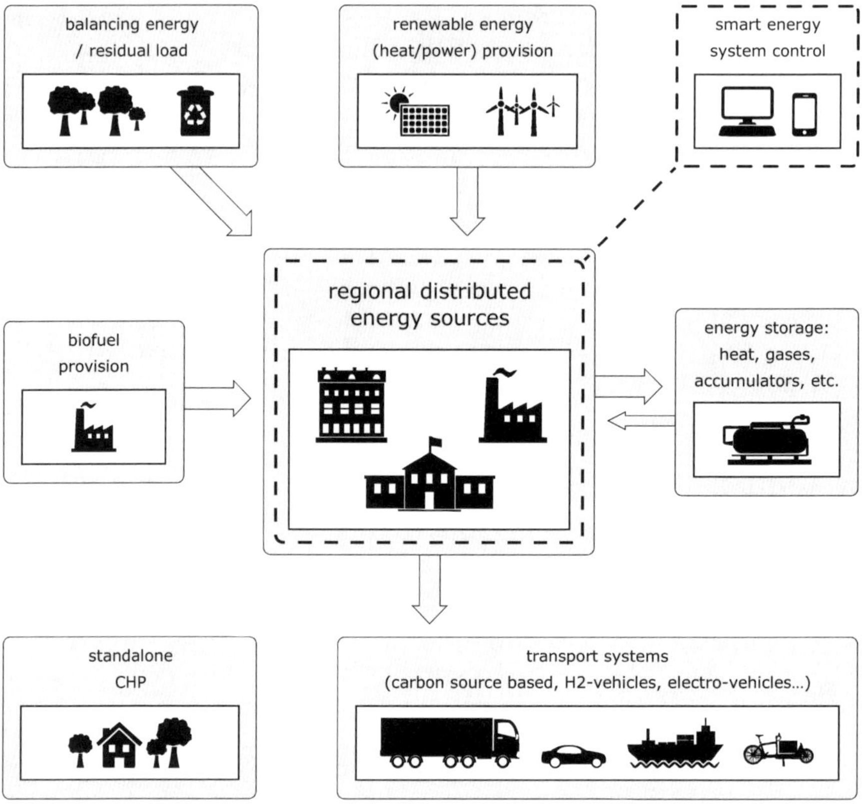

Fig. 10.2 Schematic diagram of the renewable energy system as an intelligent energy network with different bioenergy units

Even if the focus of this book was on energy use from biomass the causes and effects are particularly transferable to bio-based products: also in terms of new materials, chemicals and intermediates, the ideas and preliminary experiences with smart bioenergy could even foster decentralised and demand-driven production, which is strongly integrated into the regional resource, waste and energy flows. Such bioeconomy approaches could well develop a tremendous market potential. Flexibility is expected in terms of the feedstock, and with various products from the coupling and cascading of energy and material production. As a result, flexibility will have to be established not only between the different energy sectors, but also between the energetic and material use of biomass.

The bigger picture clearly shows three pillars for smart bioenergy systems: (i) an additional demand for smaller application units in terms of energy provision from biomass, (ii) the necessity to have improved technologies providing the desired products in small units and (iii) and new concepts of system integration – including the energy system but the coupled production of materials energy carriers from biomass as well.

Following the smart bioenergy vision the key question for bioenergy and also for bio-based materials will shift from the feedstock concerns to the development of well integrated regional concepts with smaller production units for power and heat but also smart and small biorefinery concepts as well. One relevant drive is the transition of the energy system, but considering the different frame condition in different countries it is obvious that also energy security and regional development can make the smart bioenergy approach attractive. A sustainable feedstock base, designed intermediates and controlled conversion processes in production networks are all factors that are absolutely imperative for both advanced bio-based material and energy within industrial systems based on renewable resources.

References

1. H. Chum, J. Faaij, G. Moreira, P. Berndes, Bioenergy, in *IPCC Special Report on Renewable Energy Sources and Climate Change Mitigation* (Cambridge University Press, Cambridge, UK/New York, 2011)
2. EEG, Gesetz für den Vorrang Erneuerbarer Energien (Erneuerbare-Energien-Gesetz – EEG) Konsolidierte (unverbindliche) Fassung des Gesetzestextes in der ab 1. Januar 2012 geltenden Fassung * (Renewable Energy Resources Act of Germany)
3. A. Eisentraut, A. Brown, *Heating Without Global Warming – Market Developments and Policy Considerations for Renewable Heat* (International Energy Agency, Paris, 2014)
4. Halbzeitkonferenz Bioökonomie, 5. Juni 2014. Press release retrieved from: http://www.biooekonomie.de/BIOOEKO/Navigation/DE/halbzeitkonferenz.html
5. *IPCC Review of Farming and Forests Leaves Key Questions About Effect on Climate Change "Unresolved*".http://www.carbonbrief.org/blog/2014/04/forests-and-farming-a-climate-problem,-or-a-climate-solution/ – The Carbon Brief
6. M. Junginger, C.S. Goh, A. Faaij, *International Bioenergy Trade – History, Status & Outlook on Securing Sustainable Bioenergy Supply, Demand and Markets* (Springer, Dordrecht/Heidelberg/New York/London, 2014). ISBN 978-94-007-6981-6
7. B.E. Layton, A comparision of energy densities of prevalent energy sources in units of joules per cubic meter. Int. J. Green Energy **5**(6), 438–455 (2008)
8. *Press release: Philippine Energy Minister, Carlos J. Petilla, Visits e2m Impressed by the portfolio size and technical integration.* http://www.pressebox.com/pressrelease/energy2market-gmbh/Philippine-Energy-Minister-Carlos-J-Petilla-Visits-e2m/boxid/670086v. Retrieved 7 Apr 2014
9. REN21, *Renewables 2014 Global Status Report* (REN21 Secretariat, Paris, 2014). ISBN 978-3-9815934-0-2
10. J. Rockström, W. Steffen, K. Noone, A. Persson, Planetary boundaries: exploring the safe operating space for humanity. Ecol. Soc. **14**(2), 32 (2009)
11. M. Scheftelowitz, D. Thrän, A. Krautz, J. Daniel-Gromke, V. Denysenko, K. Hillebrand, V. Lenz, J. Liebetrau, K. Naumann, u. a.: *Vorbereitung und Begleitung der Erstellung des Erfahrungsberichts 2014 gemäß § 65 EEG, Vorhaben IIa Stromerzeugung aus Biomasse* (Deutsches Biomasseforschungszentrum gemeinnützige GmbH, 2014)
12. P. Tafarte, S. Das, M. Eichhorn, D. Thrän, Small adaptations, big impacts: Options for an optimized mix of variable renewable energy sources. Energy **72**, 80–92 (2014)
13. D. Thrän, K. Bunzel, U. Seyfert, V. Zeller, M. Buchhorn, *DBFZ Report No. 7-Global and Regional Spatial Distribution of Biomass Potentials-Status Quo and Options for Specification* (DBFZ, Leipzig, 2011)
14. What is Bioeconomy? Retrieved from http://biooekonomierat.de/home-en/bioeconomy.html

Index

D. Thrän (ed.), *Smart Bioenergy*, DOI 10.1007/978-3-319-16193-8

Printed by Printforce, the Netherlands